T0212025

African Borders, Conflict, Regional and Continental Integration

This book looks at the ways African borders impact war and conflict, as well as the ways continental integration could contribute towards cooperation, peace and well-being in Africa.

African borders or borderlands can be a source of problems and opportunity. There is often a historical, geospatial and geopolitical architecture rooted in trajectories of war, conflict and instability, which could be transformed into those of peace, regional and continental integration and development. An example is the cross-border and regional response to the Boko Haram insurgency in West Africa. This book engages with cross-border forms of cooperation and opportunity in Africa. It considers initiatives and innovations which can be put in place or are already being employed on the ground, within the current regional and continental integration projects. Another important element is that of cross-border informality, which similarly provides a ready resource that, if properly harnessed and regulated, could unleash the development potential of African borders and borderlands.

Students and scholars within Geography, International Relations and Border Studies will find this book useful. It will also benefit civil society practitioners, policymakers and activists in the NGO sector interested in issues such as migration, social cohesion, citizenship and local development.

Inocent Moyo is a Senior Lecturer in the Department of Geography and Environmental Studies at the University of Zululand, South Africa.

Christopher Changwe Nshimbi is Director and Department of Science and Technology-National Research Foundation (DST-NRF) Research Fellow, Centre for the Study of Governance Innovation (GovInn), Department of Political Sciences, University of Pretoria, Pretoria, South Africa.

Border Regions Series

Series Editor: **Doris Wastl-Walter**, *University of Bern, Switzerland*

In recent years, borders have taken on an immense significance. Throughout the world they have shifted, been constructed and dismantled, and become physical barriers between socio-political ideologies. They may separate societies with very different cultures, histories, national identities or economic power, or divide people of the same ethnic or cultural identity.

As manifestations of some of the world's key political, economic, societal and cultural issues, borders and border regions have received much academic attention over the past decade. This valuable series publishes high quality research monographs and edited comparative volumes that deal with all aspects of border regions, both empirically and theoretically. It will appeal to scholars interested in border regions and geopolitical issues across the whole range of social sciences.

The Politics of Good Neighbourhood
State, Civil Society and the Enhancement of Cultural Capital in East Central Europe
Béla Filep

European Borderlands
Living with Barriers and Bridges
Edited by Elisabeth Boesen and Gregor Schnuer

Ethnicity, Gender and the Border Economy
Living in the Turkey -Georgia Borderlands
Latife Akyüz

Community, Change and Border Towns
H. Pınar Şenoğuz

Borderless Worlds for Whom?
Ethics, Moralities and Mobilities
Edited by Anssi Paasi, Eeva-Kaisa Prokkola, Jarkko Saarinen and Kaj Zimmerbauer

African Borders, Conflict, Regional and Continental Integration
Edited by Inocent Moyo and Christopher Changwe Nshimbi

For more information about this series, please visit: www.routledge.com/geography/series/ASHSER-1224

African Borders, Conflict, Regional and Continental Integration

Edited by
Inocent Moyo and Christopher Changwe Nshimbi

LONDON AND NEW YORK

First published 2019
by Routledge
4 Park Square, Milton Park, Abingdon, Oxon OX14 4RN
605 Third Avenue, New York, NY 10017

First issued in paperback 2023

Routledge is an imprint of the Taylor & Francis Group, an informa business

© 2019 selection and editorial matter, Inocent Moyo and Christopher
Changwe Nshimbi, individual chapters, the contributors

The right of Inocent Moyo and Christopher Changwe Nshimbi to be
identified as the authors of the editorial material, and of the authors for
their individual chapters, has been asserted in accordance with sections 77
and 78 of the Copyright, Designs and Patents Act 1988.

All rights reserved. No part of this book may be reprinted or reproduced or
utilised in any form or by any electronic, mechanical, or other means, now
known or hereafter invented, including photocopying and recording, or in
any information storage or retrieval system, without permission in writing
from the publishers.

Trademark notice: Product or corporate names may be trademarks or
registered trademarks, and are used only for identification and explanation
without intent to infringe.

British Library Cataloguing in Publication Data
A catalogue record for this book is available from the British Library

Library of Congress Cataloging-in-Publication Data
Names: Moyo, Inocent, editor. | Nshimbi, Christopher Changwe, editor.
Title: African borders, conflict, regional and continental integration /
[edited by] Inocent Moyo and Christopher Changwe Nshimbi.
Description: Abingdon, Oxon ; New York, NY : Routledge, [2019] |
Series: Border regions series | Includes bibliographical references and index.
Identifiers: LCCN 2018060684| ISBN 9780367174835 (hbk : alk. paper) |
ISBN 9780429057014 (ebk)
Subjects: LCSH: Regionalism--Africa. | Africa--Boundaries. |
Borderlands--Africa. | Peace-building--Africa. | African cooperation.
Classification: LCC JQ1873.5.R43 A37 2019 | DDC 327.6--dc23
LC record available at https://lccn.loc.gov/2018060684

ISBN: 978-1-03-254349-9 (pbk)
ISBN: 978-0-367-17483-5 (hbk)
ISBN: 978-0-429-05701-4 (ebk)

DOI: 10.4324/9780429057014

Publisher's Note
The publisher has gone to great lengths to ensure the quality of this reprint but
points out that some imperfections in the original copies may be apparent.

Typeset in Times New Roman
by Taylor & Francis Books

Contents

1 Borders, war and conflict in Africa
Revisiting the narrative of a war-torn continent

Inocent Moyo and Christopher Changwe Nshimbi

Introduction

The causes of conflict in Africa are many, just as they are complex. In this book we expand on this debate by focusing on the porosity of African borders and their (possible) effect in fuelling war and conflict on one hand, and the likelihood that they are sources or facilitators of peace, on the other hand. The definition of a border is paramount, because it provides an overarching frame of reference in this book. Borders are not only lines at the margins of nation-states, but are also political institutions and spaces constituted of social and political processes (Berg and van Houtum, 2003; Newman, 2006; Johnson et al., 2011; Moyo, 2016). This is aptly captured by Novak (2011: 742) who observes that borders are "both static markers of sovereign jurisdictions and socially produced and reproduced institutions". In this sense, borders perform and serve a material and symbolic function (Anderson and O'Dowd, 1999). The result of this is that they "can have a very obvious presence and even where visually indistinct, they are typically the bearers of a wider symbolism" (Anderson and O'Dowd, 1999: 595). The obvious material function and effect of borders is that they define the territoriality of nation-states. This material function and effect of borders is evident in different nation-states on the African continent and can be traced to the Berlin Conference of 1884–85. Yes, that African borders are an artificial construct occasioned by the deliberations of conveners and participants of the Berlin conference of 1884–85 is well documented (see, e.g., Miles, 2014; Bach, 1997).

The mere fact that such borders are a product of colonial gerrymandering means that these divisions are arbitrary in as much as they disregarded socioeconomic conditions and existing naturally occurring sociocultural and political delineations of the continent's inhabitants (Miles, 2014; Ramutsindela, 1999; Coleman, 1994). For this reason, people in many parts of Africa have continued to migrate from one country to the other, demonstrating that "boundary lines never proved much of a physical obstacle" (Bach, 1997: 103). This frequent interaction of people across African borders seems to suggest that the borders crossed them, rather than the people crossing the borders (Lamb, 2014; Moyo, 2016). Despite this, many so-called post-colonial states in Africa have rigidly maintained and enforced the colonial border (Moyo, 2016; Oloruntoba, 2018). This is the context within which Everisto Benyera (Chapter 2 in this volume) argues that the partitioning of Africa by the European colonisers not only enabled the imperial powers to divide and rule Africa but also planted the seeds of conflict. It also forced African countries "to enter into the global capitalist, legal and Euro-North American moral order" and also bequeathed postcolonial states with a lot of lethal institutions such as borders (see, Chapter 2). The extent to which these institutions are lethal is evidenced by, among others, the fact that the borders divide rather than unite African people (see, e.g., Flynn, 1997; Moyo, 2016; Nshimbi, 2017). It can also be illustrated by the way in which most African countries approach the movement of people across these institutions in contiguous border areas which separate people who share common histories, ethnic identities and cultures. Flynn (1997), Moyo (2016) and Nshimbi (2017), respectively study borders in countries that span West, Southeast and Southern Africa. They each tell of the ways in which people who dwell in the borderlands of disparate African countries in those places stake their claim to cross-border movements; sometimes defying state authorities because, as far as these people are concerned, the border is an institution imposed on and dividing their longstanding habitat and communities. Such people are referred to as border citizens (Meeks, 2007; Moyo, 2016). But today's African state, demarcated almost a century and half ago by colonialists, and modelled according to the Westphalian state designed in the global North, apparently opts to ignore such historical and contemporary ethnocultural and social realities. The African state chooses instead to treat as law breakers, people who are separated by the social and politically constructed symbols which ignore the fact that these peoples are drawn together across such lines by strong historical and filial bonds. This has led to the criminalisation of migration between African states and the resultant xenophobia, as demonstrated in Chapter 8 of this volume.

With this in mind, the chapters in this volume engage with the coloniality of African borders; and the extent to which such coloniality led to the balkanisation of African countries and the conflict that this engendered (Chapter 2) in some parts of Africa. These borders are abyssal lines. They have constructed "others" out of people who have always been one community. Because of the balkanisation of African countries, it is no wonder that their borders are porous and serve as theatres of cross-border and regional conflicts for some countries (see Chapter 3, Chapter 4, Chapter 5 and Chapter 6 in this volume).

There are various symbolic functions of borders – one of which may be "putting distance in proximity" (Groupe Frontière et al., 2004 cited in Szary, 2015: 36) – suggesting that borders are "a kind of space where the relationship with otherness can be developed in such a way as to allow for identity-building and place-making" (Szary, 2015: 36). Just like in their material forms, as social and political institutions, borders have been used as processes to create and increase distance between and amongst African people, thus

engendering xenophobia. Further, they have been used to create an architecture which provides a foundation for and thus sustain a trajectory of violence (see Chapter 3, Chapter 4, Chapter 5, Chapter 6, Chapter 8 and
Chapter 9 in this volume).

From the foregoing, it is evident that borders as material and symbolic
institutions have a divisive impact on African countries. However, we are also
using the same logic of the materiality and symbolism of borders to advance
the view that, beyond fostering a trajectory of conflict, violence and war,
borders can also provide an architecture for peace in Africa. This is the logic
behind regional integration schemes and the ultimate unification and "elimination" of borders in Africa, as immortalised in the 1991 Abuja Treaty for
Establishing the African Economic Community and the Agenda 2063 project.
It is the context within which we advance the innovation that, as social or
symbolic institutions, borders can be used to promote peace and integration
on the continent. If borders can be used to put "distance in proximity"
(Groupe Frontière et al., 2004 cited in Szary, 2015: 36), they could equally be
used to move distance to proximity between and amongst African people.
This is because there are networks and different sets of cross-border archi-
tectures that existed in precolonial and survived colonial and post-colonial
systems. Such can be easily utilised for the purpose of development, peace,
integration and African unity (see Chapter 3, Chapter 7). By suggesting that
African borders can be innovatively used for establishing peace in Africa, we
are not oblivious to the presence of other processes and/or factors within
African states and, indeed, other "borders" that can militate against peace.
But this is a modest attempt at initiating debate which will ultimately lead to
a conceptualisation of African borders and borderlands beyond zones and
spaces where chaos and violence reign supreme. In this vein, the different
contributions in this volume attempt to bring to the fore the different angles
to this debate.

The reality of the African border

Thus, the liberal movements that occur across Africa's extensive and porous
borders, which in some cases disregard the colonial and post-colonial borders,
do not signify either the non-existence or the reality of the border. In fact,
African borders matter. They have an impact to the extent that they "separate
areas in war from countries at peace and democracies from dictatorships; they
structure settlement patterns, cultural forms, typical economic activities, consumption opportunities or kinship ties" (Dobler, 2016: 147). This means that,
different administrative mechanisms, among other factors, account for why
some countries prosper and are peaceful, while others do not enjoy these
positives (Dobler, 2016). These differences emphasise that borders do matter
(Dobler, 2016). It is these differences between African states that directly or
indirectly trigger cross-border movement. This, in essence means that cross-
border movement is inevitable and that it occurs within the context where

some African states obsess about strengthening the border. But the strengthening of porous African borders actually does not deter and may, in fact, generate other forms of cross-border migration. Cross-border movements may occur at undesignated crossing points. This is pertinent in this age, when some undesirable phenomena such as terrorism seen in the activities of Boko Haram in West Africa or Al Shabab in the Horn of Africa, for example, have been on the increase.

The African border as a source of conflict

The colonial border in Africa is, itself, a (potential) source of conflict too. This arises from attempts by some of the continent's post-colonial states to enact or redefine the colonial border they share with fellow post-colonial states. Tensions of this sort are spread out across Africa, with associated challenges including terrorism, ethnic violence, military clashes, cattle rustling, smuggling of goods as well as human smuggling and human trafficking, among others. Multiple challenges underlying border disputes in this regard fundamentally boil down to claims over or access to natural resources. In Southern Africa, for example, South Africa and Namibia have long contested over the Orange River while Botswana has had its own disagreements with Namibia over the exploitation of the Okavango River (Nshimbi, 2017). In North Africa, Egypt administers the Halaib region, which Sudan claims while the Western Sahara – that former Spanish territory annexed by Morocco – remains at the centre of disagreements among Morocco, Mauritania, Algeria and Libya. In East Africa, Kenya and Sudan dispute over the Ilemi Triangle while disputes over territories such as the Badme characterise relations between Eritrea and Ethiopia. And in Central Africa, the Congo River, flowing through the Republic of Congo (Brazzaville) and the Democratic Republic of the Congo (DRC), is a source of dispute between the two countries over the location of the border between them, within the water body.

The foregoing being the case, this book examines borders in Africa and how their porosity potentially contributes to the exacerbation of war and conflict or to fostering cooperation, peace and well-being. Questions concerning the ways in which the porosity of African borders could catalyse instability and the spread thereof from one country to several others, as in the case of Nigeria, where insurgency and counter insurgency has spread to parts of the Economic Community of West African States (ECOWAS) region and Central Africa, arise and are examined in this book. Indeed, the contributions by James Okolie-Osemene and Benjamin Adeniran Aluko in this volume (Chapter 4) and Nelson Alusala (Chapter 5) clearly demonstrate that, the porosity of borders in West Africa and the DRC and the Great Lakes Region exacerbated cross-border conflict and instability. In the latter, the porosity of the borders, especially between the DRC and Rwanda, has promoted the merging of proximate and core causes of conflict in the DRC. This has led to the exacerbation of the DRC's internal conflict, ultimately transforming it into a regional one. In this sense, and as a result of the porosity of borders, instability in one country or society can contagiously spread across the border and catalyse events in the neighbourhood, into an active conflict.

Jackson Aluede (Chapter 6), takes up this argument based on a comparative analysis of the cross-border dimension of intra-state conflicts in the Great Lakes (Burundi, Rwanda and the DRC) and the Mano River (Liberia and Sierra Leone) regions. By employing theories of protracted social conflict (PSC), transnationalism and ethnicity, Jackson argues that the porous nature of the borders, coupled with other inherent challenges confronting the states – ethnic divisions, forced migration, bad governance, rebel incursion and others – contributed to the cross-border dimension of the conflicts (Chapter 6). Although intra-state conflicts in the Manor River ended around 2003, the author concludes the discussion by highlighting that, both the sub-national and national apparatuses of states in both regions could use the porous borders as a vehicle to achieve peace in the region. In the latter, the porosity of borders has complicated efforts targeted at the Boko Haram insurgency in Nigeria and the neighbouring West African states. Through the deployment of the political theory of territory and that of ungoverned spaces, the authors assert that, as a result of the porosity of borders between Nigeria and its neighbouring states, insurgency has proved difficult to contain despite border security efforts. At the same time, these authors observe that the same porous borders can also be mobilised towards counter insurgency efforts through the prioritisation of trans-border security collaboration and the timely incorporation of insurgence emergency response systems and border security governance with active engagement of communities by the states concerned. In other words, the porosity of borders should not only be seen in a negative light, in terms of the efforts aimed at responding to the Boko Haram threat.

Borders, co-operation, peace and development?

Also examined are questions about the innovative ways in which some African countries opt to cooperate in exploiting natural resources at their common borders in pursuit of common political goals, which also leads to peace (Nshimbi, 2017). This is predicated on the existing strategy for continental integration. Africa's strategy for attaining integration by establishing the African Economic Community (AEC) is, indeed, through Regional Economic Communities (RECs). These should provide the foundation for integration based on eight such RECs on the continent; to move Africa towards the AEC.[1] The establishment of the African Continental Free Trade Area (AfCFTA) and the AEC resonates with the African Union's (AU) commitment to speeding up actions that will "silence the guns by 2020", in the AU's Agenda 2063. This is in order to make peace a reality for all people in Africa (Agenda 2063: 6). This means that peace is linked to security and development (Murithi, 2011), and the establishment of the AfCFTA and the AEC are aspects of development, whose success may not be possible outside peace in Africa. So, do wars or conflicts militate against the success of RECs and the ultimate integration of the African continent? This is a pertinent question to ask because the 1991 Abuja Treaty for Establishing the African Economic Community clearly outlines the AU's determination to establish an economic community of African states by 2028, preceded by the AfCFTA. The AfCFTA agreement has since been signed: at the AU Special Summit held in Kigali, Rwanda on 21 March 2018.

The question of whether wars or conflict affect regional integration is tackled by Ntongwa Bundala (Chapter 10) in this book. Ntongwa argues that although war and conflict are the least factors that cause the failure of the African regional integration plans, the presence of peace equally creates a good environment for regional integration. The implication of this is that peace is good for regional integration. In turn, regional integration can promote Africa because "RECs increase economic and political trust among African countries" (see, Bundala, Chapter 10 in this volume). But there is need to critically understand the concept of peace, beyond just the absence of war within the setting of regional and continental integration. In this regard Inocent Moyo (Chapter 8), asserts that integral to the process and implementation of regional and continental integration is the issue of human mobility. Inocent argues that African countries, particularly those in the Southern African Development Community (SADC) should devise migration policies which promote the movement of people, as stringent migration policies are responsible for the uncertainties and anxieties of migrants who move within the region, such as those from Zimbabwe to South Africa. Uncertainties and anxieties of migrants create negative peace, which is comparable in its effects on migrants to people who are fleeing actual war and conflict. For this reason, Inocent argues that in line with the regional integration process in the SADC, there is need for a reconfiguration of immigration policies which reflect the reality of integration in the form of free human mobility. This free human mobility is an essential element of regional peace, if peace is understood as not only the absence of war, but also the absence of other intervening factors and structures that disturb the inner peace and security of people (Francis, 2011). Stringent immigration regimes, which place people in a tenuous immigration status, are a good example of such. This actually shows that states, especially those in the SADC region, abandon the ordinary people in pursuit of statist and colonial bordering policies.

Concerning the SADC states abandoning its people, Patrick Dzimiri (Chapter 9) invokes the case of the 2008 disputed elections in Zimbabwe to advance the argument that the region endorsed a dictator, in the person of President Robert Mugabe, at the expense of the Zimbabwean people. The SADC states did this in the name of solidarity and regional integration at the level of the state, but ordinary citizens experienced untold suffering which led to massive migration from Zimbabwe to neighbouring countries like South Africa, Botswana, Namibia and Zambia. In some of these countries, Zimbabweans were met with virulent xenophobic attacks in which some lost their lives. Patrick posits that, although at the level of the state there was no war following the 2008 disputed elections in Zimbabwe, at the grassroots level it was a different scenario.

In this vein, Patrick states that:

> it is fair to argue that SADC states wanted to protect a dictator for the sake of championing fake Afro-radicalism ... and the defence of independence against the so-called vampirism of neo-colonial matrices of power ... . Further examination of the human security costs of the electoral conflict [in Zimbabwe] points to the fact that, [the so called] regional integration (defined [by Patrick] as the club of people/leaders who fought against colonialism) should exist at the expense of the peace of the ordinary person, as long as the SADC had prevailed against the Western threat – real or imagined [by siding with and supporting Mugabe].
>
> (see Chapter 9)

Therefore, states should consider the needs of the ordinary person on the ground both in terms of the policies that they enact and implement as well as in simple processes such as elections.

Elections in the SADC region have a bearing on borders, migration, regional integration and peace in the sense that, if they are not properly monitored, as the case of Zimbabwe shows, they lead to the displacement of people. This can destabilise regional peace. It suggests that the peace that the region enjoys at the level of state-to-state relations may be fake, as long as people on the ground are displaced and live under difficult conditions largely created by those at the state level. But, this can all be solved if African countries, such as SADC member states, seriously consider the needs of the people and enact robust regional systems ranging from efficient election security policies to managing cross-border migration.

Managing transboundary natural resources for development

If porous borders could precipitate or be a source of conflict, or, if they constitute the means for establishing peace, could natural resources that cross nation-state borders and are shared by at least two countries, such as transboundary waterbodies, be harnessed in order to be of benefit to all who share the resources? In what ways could the resources be governed in order to particularly benefit non-state actors who inhabit the borderlands of the proximate states that share the resources? Christopher Nshimbi (Chapter 7) attempts to answer these questions through a policy analysis of measures designed to regulate the use of natural resources; particularly the effects on grassroots actors whose sociocultural being is subjected to the division brought about by the arbitrary construction of a line of separation in the form of a political border, and whose livelihoods depend on the shared transboundary resources. Citing the activities of artisanal fishers on a transboundary water body shared by Zambia and Zimbabwe, Chris highlights the importance of the activities of these actors for employment and the sustenance of livelihoods.

Also important is the fact that the activities on the shared transboundary water body highlight tensions and conflict between grassroots actors and the state, and the neglected aspect of peaceful coexistence of actors in the borderlands communities. Grassroots actors who traverse borders in the SADC region secure livelihoods as they go about their activities and operations. This means that their families are supported and food security is achieved – both very important elements of peace, if it is understood to be not the mere absence of war. For this cause, Chris proposes that the states, together with local governing authorities need to proactively establish governance structures for the management of cross-border resources and the activities of grassroots cross-border actors. This will enhance the management of the shared resources and contribute to the achievement of the United Nations (UN) sustainable development goals (SDGS), sustain livelihoods, and contribute to the bottom-up processes of regional integration and, ultimately, peace.

A related case is that of the Afar Horn, which is constituted of Ethiopia, Djibouti and Eritrea. Muauz Gidey (Chapter 3), asserts that the countries in this region share a common history, identities, geospatial communities as well as economic and other ties. For this reason, it is not surprising that the region is entangled in conflict, insecurity and similar development challenges. However, the borders, history and assets that characterise the region should not only be seen as raw materials that fuel conflict, war and instability, but could be harnessed by states so as to unleash the development potential of the region. Muauz asserts that, in this sense, the existing historical, geospatial and geopolitical architecture, which engenders a trajectory of war, conflict and instability across the borders in the region, could be transformed to that of development, peace and prosperity. In this context, borders and borderlands in the Afar region can be used as an architecture for achieving peace in Africa.

Conclusion

For the reasons cited in the foregoing, beyond the question of how the porosity of African borders exacerbate conflict and war, this book engages with the question of the extent to which the porosity of African borders can also be a source of regional peace. This volume thus engages with the issue of African borders and how the borders can either fuel war and conflict or become an instrument for forging cooperation and peace. If African borders can be utilised in this way, the issue of regional and continental integration in Africa through RECs comes into sharp focus and is thus explored in this book.

Note

1 The eight RECs are the Community of Sahel-Saharan States (CEN-SAD), Common Market for Eastern and Southern Africa (COMESA), East African Community (EAC), Economic Community of Central African States (ECCAS), Economic Community of West African States (ECOWAS), Intergovernmental Authority for Development (IGAD), Southern African Development Community (SADC), Union du Maghreb Arabe (UMA). See www.au.int/en/recs/. [Accessed 26 November 2012].

References

Abuja Treaty establishing the African Economic Community (AEC). 1991. https://pm g.org.za/committee-meeting/243/ (accessed March 2017).

Agenda 2063. 2015. *Agenda 2063: The Africa We Want*. https://au.int/sites/default/file s/pages/3657-file-agenda2063_popular_version_en.pdf (accessed 10 May 2018).

Anderson, J., and O'Dowd, L. 1999. Borders, Border Regions and Territoriality: Contradictory Meanings, Changing Significance. *Regional Studies* 33(7): 593–604.

Bach, D. 1997. Frontiers Versus Boundary Lines: Changing Patterns of State–Society Interactions. *Welt Trends* 14: 97–111.

Berg, E. and Van Houtum, H. 2003. *Routing Borders between Territories: Discourses and Practices*. Burlington, VA: Ashgate.

Coleman, J. S. 1994. *Nationalism and Development in Africa*. Berkeley: University of California Press.

Dobler, G. 2016. The Green, the Grey and the Blue: A Typology of Cross-Border Trade in Africa. *The Journal of Modern African Studies* 54: 145–169.

Francis, T. 2011. Linking Peace, Security and Developmental Regionalism: Regional Economic and Security Integration in Africa. In Erin McCandless and Tony Karbo (Eds), *Peace, Conflict and Development in Africa: A Reader*. Addis Ababa: University for Peace, Africa Programme, pp. 210–520.

Flynn, D. K. 1997. "We Are the Border": Identity, Exchange, and the State along the Benin–Nigeria Border. *American Ethnologist* 24(2): 311–330.

JohnsonC., Jones, R., PaasiA., Amoore, L., Mountz, A., Salter, M., and RumfordC. (2011). Interventions on Rethinking the Border in Border Studies. *Political Geography* 30: 61–69.

Lamb, V. 2014. "Where is the Border?" Villagers, Environmental Consultants and the Work of the Thai–Burma Border. *Political Geography* 40: 1–12.

Meeks, E. V. 2007. *Border Citizens: The Making of Indians, Mexicans, and Anglos in Arizona*. Austin: University of Texas Press.

Miles, W. F. S. 2014. *Scars of Partition: Postcolonial Legacies in French and British Borderlands*. Lincoln: University of Nebraska Press.

Moyo, I. 2016. The Beitbridge–Mussina Interface: Towards Flexible Citizenship, Sovereignty and Territoriality at the Border. *Journal of Borderlands Studies* 31(4): 427–440.

Murithi, T. 2011. African Institutions: Securing Peace and Development across Borders. In Erin McCandless and Tony Karbo (Eds), *Peace, Conflict and Development in Africa: A Reader*. Addis Ababa: University for Peace, Africa Programme, chapter 11.Newman, D. 2006. Borders and Bordering: Towards an Interdisciplinary Dialogue. *European Journal of Social Theory* 9(2): 171–186.

Novak, P. 2011. The Flexible Territoriality of Borders. *Geopolitics* 16: 741–767.

Nshimbi, C. C. 2017. The Human Side of Regions: Informal Cross-border Traders in the Zambia–Malawi–Mozambique Growth Triangle and Prospects for Integrating Southern Africa. *Journal of Borderlands Studies*, DOI: doi:10.1080/08865655.2017.1390689.

Oloruntoba, S. O. 2018. Crisis of Identity and Xenophobia in Africa: The Imperative of a Pan-African Thought Liberation. In O. Adeoye Akinola (Ed.), *The Political Economy of Xenophobia in Africa*. Cham, Switzerland: Springer International Publishing, pp. 9–22.

Ramutsindela, M. 1999. African Boundaries and Their Interpreters. *Geopolitics*, 4(2): 180–198.

Szary, A.-L. A. 2015. Boundaries and Borders. In J. Agnew, V. Mamadouh, A. J. Secor and J. Sharp (Eds), *The Wiley Blackwell Companion to Political Geography*. Chichester: John Wiley and Sons, pp. 25–50.

2 Borders and the coloniality of human mobility

A view from Africa

Everisto Benyera

Introduction

Borders perform many functions, such as separating countries at war from those at peace, and democracies from dictatorships; they structure settlement patterns, cultural forms, typical economic activities, consumption opportunities or kinship ties, among others (Dobler, 2016). This notwithstanding, borders are not innocent infrastructure, especially for those in (formerly) colonised parts of the world, and especially Africa, which is the focus of this chapter. Concerning African borders, a background is needful. About two centuries before the Berlin Conference of 1884–85, after the 30 years' war in Europe, Europeans gathered in Westphalia, Germany, and agreed to thenceforth respect each other's sovereignty. That was the birth of the cherished notion of state sovereignty. Ironically, it was also at the Berlin Conference that European leaders of the day agreed to arbitrarily partition Africa, a development well addressed in scholarship (Grosfoguel and Cervantes-Rodriguez, 2002; Ndlovu-Gatsheni, 2015; Nimako, 2015; Táíwò, 2009). While the reasoning of the delegates at the signing of the Treaty of Westphalia and the Berlin Conference, respectively, to respect each other's state sovereignty and to deny Africa and Africans their state sovereignty, is illogical, it is consistent with imperialist colonial expansion. This is best illustrated by the facts that:

> First, colonialism and capitalism forcibly incorporated Africa into the world economy, beginning with the slave trade … . Secondly, Africa had been excluded from the post-1648 Westphalian sovereign state system … Third, Africa was incorporated into a Euro-North American-centric world culture and European languages. Fourth, Africa was incorporated into a heavily Euro-North American-centric world of international law. Fifth, as a consequence of colonialism, Africa was incorporated into the modern technological age including being "swallowed by the global system of dissemination of information". Finally, Africa was dragged into a Euro-North American-centric moral order dominated by Christian thought.
>
> (Ndlovu-Gatsheni, 2015: 47–48)

The resilience and perseverance of the asymmetrical colonial power relations between the (formerly) colonised peoples and their (former) colonisers is termed coloniality. It is coloniality which defines the way borders between the (formerly) colonised and the (former) colonisers are operated (Ndlovu-Gatsheni, 2013). Coloniality differs from decolonisation in that, decolonisation denotes the departure of the colonisers from the colony back to the metropolis. But they left behind a sophisticated system and network of infrastructure which allows them to dominate these relations (Ndlovu-Gatsheni, 2013). Within this system, colonial borders as institutions bequeathed on (post-)colonial states by colonialism, allow the (former) colonisers to enjoy what is termed distributed presence in their (former) colonies.

Arguing from a decolonial perspective, Ndlovu-Gatsheni (2015) views borders in Africa as a product of coloniality of power and contends that coloniality of power is still productive in policing western borders so that, *inter alia*, those from the (former) colonies do not enjoy unfettered access into the metropolis. Coloniality, indeed, denotes the resultant relationship in which power relations remain tilted heavily in favour of the (former) colonisers, even after the official end of colonialism. Ndlovu-Gatsheni argues:

> The concept of coloniality of power enables delving deeper into how the world was bifurcated into "Zone of Being" (the world of those in charge of global power structures and beneficiaries of modernity) and "Zone of Non-Being" (the invented world that was the source of slaves and victims of imperialism, colonialism, and apartheid) maintained by an invisible line that Boaventura de Sousa Santos termed "abyssal thinking".
>
> (Ndlovu-Gatsheni, 2015: 490)

Once the zone of non-being was officially created at the Berlin Conference, the colonisers started engaging in concerted efforts to divide and rule the people occupying this zone through the construction of ethnic differences and identities, in some cases with the complicity of the Church (Ranger, 1989). How the Belgian colonisers constructed the ethnic identities of the Hutus and Tutsis in Burundi and then weaponised them, is a well-documented case which largely contributed to the genocide of 1994 (Clark, 2007; Mamdani, 2010; Ndlovu-Gatsheni, 2010: 289; Senier, 2008). Based on such differences, instances exist in which people in the zone of non-being do not enjoy social cohesion and, even, engage in conflict against each other on the basis of ethnic difference.

Mahmood Mamdani differs from those whom he terms Africanists who blame Euro-North American modernity for the "border challenges" Africa faces. Mamdani writes:

> Well, what would be genuine boundaries? From this point of view, the answer would be that they would be "natural," meaning they would not cut through ethnic boundaries. In other words, the political map of Africa

should have followed its cultural map. I find two problems with this kind of argument. All boundaries are artificial; none are natural. War and conquest have always been integral to state-building ... The real problem with this point of view is the assumption that cultural and political boundaries should coincide, and that the state should be a nation-state – that the natural boundaries of a state are those of a common cultural community.

(Mamdani, 2015: 4)

A counter-argument to Mamdani would be: what is wrong with a common cultural community as the basis for nation-states, because in Mamdani's words African borders, "were drawn up with a pencil and a ruler on a map at a conference table in Berlin in the 1880s" (Mamdani, 2015: 4)? Various forms of violence were then used to enforce and institutionalise these borders. Forcing (former) colonies into the global capitalist, legal and Euro-North American moral order (Mazrui, 1986; Ndlovu-Gatsheni, 2015: 487–488), through what Achille Mbembe termed foundational and maintenance violence (Mbembe, 2000), locked (former) colonies into various forms of borders that preserve the colonial *status quo*. Explaining Mbembe's three forms of colonial violence which were efficacious in the normalisation of colonial order, Benyera writes:

> [There was] "foundational violence" which authorised the right of conquest and had an "instituting function" of creating Africans as its targets; ... "legitimating violence," was used after conquest to construct the colonial order and routinise colonial reality; and "maintenance violence," was infused into colonial institutions and cultures and used to ensure their perpetuation.
>
> (Benyera, 2017: 69)

These three forms of violence, where the imperialists laid the foundation for colonial borders in Berlin, used law, war, religion, morality, commerce, culture and language to legitimise them and then used institutions and structure to maintain the *nefariousness* of borders in Africa today. Another way of analysing African borders and their effects is to use what is termed the locus of enunciation (Mignolo, 2009). The notion of locus enunciation states that the same phenomenon can be interpreted differently depending on where one is located (epistemologically and geographically). Mignolo argues:

> Westerners then defined the locus of enunciation (not as geo-historically and geo-politically located, but as the enunciation of the universal). Easterners defined instead the enunciated to whom the enunciation was denied: Easterners were defined by Westerners as if Westerners had the universal authority to name without being named in return.
>
> (Mignolo, 2009: 8)

Hence others looking at African borders from the West and a western perspective naturally arrive at different conclusions from those analysing borders
from the perspective of (those who were previously) the colonised. Those in
what is loosely termed the global South are not contesting the authenticity of
their views. Rather they are presenting an alternative perspective on borders,
one conceptualised from what Fanon termed the zone of non-being. According to Fanon, "there is a zone of non-being, an extraordinarily sterile and
arid region, an utterly naked declivity where an authentic upheaval can be
born" (Fanon, 1952: 2). This zone is the place where those in the global
South analyse borders from.

In arguing for the use of locus of enunciation as a decolonial perspective to
provide an alternative view of "the border", for instance, William Mpofu
argues that the locus of enunciation:

> is a concept that is founded on the thinking that in a "pluriversal
> world", and a world that espouses not just one imperial knowledge, but
> "ecologies of knowledge" all knowledges are situated and located. In
> arguing for "shifting the geography of reason in the age of disciplinary
> decadence", Afro-American philosopher of humanity, Lewis R Gordon
> depicts the geopolitics of knowledge, where knowing and thinking are
> grounded. In the same way, when Paula ML Moya insists on the
> importance of being "who we are, from where we speak" she argues for
> a geo-politics and a body-politics of thinking and knowing from an
> announced locale.
>
> (Mpofu, 2014: 15–16)

As a result, in this chapter, I carry forward the context provided in Chapter 1
and ask the following questions: what are borders? What is their function?
What is the result of the creation of borders? Who has the authority to (re)
draw borders? And, what processes were used to (re)draw borders in Africa,
at least? Because the first two questions have been addressed in Chapter 1, I
will focus more on the last three and make them the chapter's central research
contribution. I will answer them in the next sections, and use Africa and
Euro-North America as my reference cases; drawing especially on their colonial and post-colonial relationships. Looking at African society and economies today, we realise how relevant borders have become over the last one-
and-a-half centuries, as they (colonial borders) are increasingly being unchal-
lenged by (post-)colonial African governments.

Conceptualising borders as abyssal lines

As Moyo and Nshimbi (see Chapter 1) have conceptualised borders, I also
argue that borders are not only a physical phenomenon but have many facets,
forms and functions. I rely on the notion of the abyssal line, which was postulated by Bonaventura de Sousa Santos (2007), in developing this section.

For de Sousa Santos, coloniality operates through enacting abyssal lines, which are constituted "of a system of visible and invisible distinctions" (de Sousa Santos, 2007: 45). When conceptualised as abyssal lines which divide the zone of being from the zone of non-being, it becomes clear that borders are not only geographical in nature but assume a variety of other forms which give rise to terms such as "illegal" immigrant and "undocumented" immigrant. Accordingly, a human being becomes "illegal" the moment they leave "their native land" and cross the abyssal line in flight from, among other things, deprivation, persecution, war and conflict. As the illegal or undocumented immigrants try to fit into the zone of being, the "border" pursues them, as it refuses to be left at the physical border (Moyo and Nshimbi, 2017). So omnipresent is the border that those deemed to be illegal immigrants are routinely rounded up and deported.

The argument that borders have evolved into institutions and mechanisms as opposed to simple lines at the margins of nation-states has been sufficiently advanced (Cons and Sanyal, 2013; Laine, 2015; Novak, 2011; Paasi, 2012), and this work is hereby acknowledged for laying the foundation on which the main arguments of this chapter are premised. As argued by Laine:

> Today, borders are widely recognised as complex multileveled and -layered social phenomena related to the fundamental organisation of society as well as human psychology. This has not, however, been always the case, but the way borders have been viewed and interpreted has evolved – much in line with broader discursive shifts in social sciences as well as in relation to overlying geopolitical events.
>
> (Laine, 2015: 14)

In extreme cases borders consist of physical walls erected between two countries, the obvious case being the Israel–Palestine wall, and the one proposed at the Mexico–USA border. In this chapter, I argue that the border is more than a physical phenomenon, but one which consists of both visible and invisible lines. "The invisible lines form the foundation for visible ones" (de Sousa Santos, 2007: 45). Borders form a painful reality for (previously) colonised people. As de Sousa Santos argues, abyssal lines give rise to two contrasting situations "this side of the line" and "the other side of the line" (de Sousa Santos, 2007: 54) where life on the two different sides is the exact opposite of the other in terms of people's existentialism. Borrowing from Fanon's (1952: 2) zones of being and non-being seen earlier, adverse conditions in the zone of non-being explain the many perilous journeys across the Mediterranean Sea taken by Africans seeking to cross the abyssal line and enter the zone of being, which is Euro-North America. When the border is conceptualised as an abyssal line, we are then able to make sense of how this institution works in metropolitan societies as well as and in contrast to colonial territories. In particular, the border in the zone of non-being was used as an instrument for subjugating the colonised. This was because the colonisers also engaged in a concerted effort to divide and rule the people within this zone, through the construction of ethnic differences and identities. Based on such difference, instances exist in which people in the zone of non-being are not socially cohesive, and even war against each other on the basis of the colonially constructed ethnic difference – wherein processes of othering and construction of "us" and "them" occur. This is the context within which border and intra-state conflicts on the African continent are ignited and spread. Such conflicts involve essentially the same people, who were nonetheless reconstructed and misrepresented as ethnically different by the colonial project (see, e.g., Chapters 4 and 5 in this volume).

Types of borders

The obvious types of borders are the national, provincial and other administrative borders which geographically demarcate jurisdictions. There are many other latent forms, such as the structural, ontological and epistemological borders. By ontological borders, I am referring to the system of ascribing ontological density and validity to humanity mainly based on their place of birth, gender and race (Mignolo, 2009). In this system, the Euro-North American white male has more ontological density than an African black female (Mignolo, 2009) and on the basis of this, the "modern/colonial capitalist/patriarchal western-centric/Christian-centric world-system has privileged the culture, knowledge, and epistemology produced by the West inferiorising the rest" (Grosfoguel, 2011: 10).

Then there are structural borders, which relate to systemic issues, which affect black people. For instance, when a black person graduates from the *anthropos* to join the *humunitas* (Benyera, Mtapuri and Nhemachena, 2018), they are still criminalised and denied entry into the system primarily because of their race. The racial border is described by Chinweizu thus:

> if you are white and running down the street, you are an athlete; but if you are black and running down the street, you are a thief! And in most parts of the world today, if you are white and rich, you are honoured and celebrated, and all doors fly open as you approach; but if you are black and rich, you are under suspicion, and handcuffs and guard dogs stand ready to take you away. Yes, the black skin is still the badge of contempt in the world today, as it has been for nearly 2,000 years.
>
> (Chinweizu, 1993: 1)

The result of this ontological ordering of humans is that movement from one layer to another is prevented by a solid "border" consisting of privileges, norms, values and standards whose combination renders it almost impossible for black mankind to enter the realms of white mankind (Ndlovu-Gatsheni, 2013). Similarly, epistemological borders consist of a system which orders knowledge, in the process privileging Euro-North American knowledge. Not only is Euro-North American knowledge privileged, it is also universalised, taken as the standard and rarely questioned. Knowledge from elsewhere in the (previously) colonised world is often doubted and treated as myth (Ndlovu-Gatsheni, 2013). The opening of colleges of the University of London in colonies of the United Kingdom actualised this system of exporting the British education system to colonies. Once planted and operational, (former) colonies are today stuck with a colonial education resulting in what has been termed the western university in Africa among other institutions. Indigenous knowledge and knowledge systems are locked outside the "borders" of knowledge (Ndlovu-Gatsheni, 2013).

Other types of borders are social, cultural, economic, class, residential, health, educational, gender, occupational, professional, associational, organisational and generational (Benyera, 2018: 142; Mtapuri, Nhemachena and Benyera, 2018). Movement from one culture to another or trying to enter a culture is not very difficult in Africa because of the precolonial and colonial legacy of African states such as the Ndebele, Torwa, Mutapa and Mapungubwe which used to accept citizens from other nations as part of their nation building processes (Cobbing, 1974). Ndlovu-Gatsheni argues this notion well when he states that:

> Precolonial African socio-political formations had room for full incorporation and successful assimilation of defeated communities into the host society. But under colonial modernity that was shot through with a racial order of identities, whites could not be accommodated into the African societies they despised and sought to transform and black people could not be accommodated into colonial white society that was fenced in by racism.
>
> (Ndlovu-Gatsheni, 2013: 127)

Unlike colonial borders which are rigid, there are ways of gaining cultural acceptability such as performing certain rites and rituals. However, with Euro-North American cultures, one can enter through ways such as *assimilado*, but equal status will never be granted as such communities remain structured according to racism. For Ramón Grosfoguel, racism has become an instrument of ordering humans according to ascribed superiority/inferiority from which one can hardly escape. Grosfoguel argues:

> Racism is a global hierarchy of superiority and inferiority along the line of the human that have been politically, culturally and economically produced and reproduced for centuries by the institutions of the "capitalist/patriarchal western-centric/Christian-centric modern/colonial world-system".
>
> (Grosfoguel, 2015: 10)

Economic and class "borders", while theoretically permeable, are difficult as those who possess land and generational wealth tend to acquire more wealth while those born in poverty tend to stagnate or regress into more poverty. The intersection of race, class and economic status makes this "border" harder to permeate, in the process resulting in racial privilege and its opposite, racial prejudice (Grosfoguel, 2015: 10–11). Those endowed with racial privilege are viewed and treated as tourists and/or investors when they cross borders, while those experiencing racial prejudice are treated as perpetual criminals, suspects, crimes waiting to happen (Chinweizu, 1993).

Health "borders" are more visible when analysed through the private/ public hospital dichotomy. In countries like South Africa, those who can afford medical aid get world class health care while the rest rely on public facilities which are overloaded and lacking in capacity (Statistics South Africa, 2015: 73). Getting medical aid equates to crossing the abyssal line and is literally the difference between life and death. Then, there are associational "borders" where in order to belong and be a member of certain powerful network organisations and associations, one has to be recommended by those within. This ensures that "outsiders" are kept out and the association remains "original". It is at these "croquet and tennis" associations that decisions affecting whole communities and countries are made.

Generational "borders" imply those differences where the young and the old have different perspectives on the same phenomenon. The recent Brexit referendum in the United Kingdom was won and lost largely due to differences in generational interpretations of belonging and nationalism (Anon, n. d.; Curtice, 2016; Hobolt, 2016; McEwen, 2016). There are many forms and types of borders whose exploration and analysis warrants more attention than can be afforded in this chapter. That aside, my focus here is on the material and symbolic functions of borders between and within African countries.

I conclude this section by briefly discussing the notion of "gender borders" as the many restrictions placed on women which emanate from patriarchal, masculine, sexist, colonial, heteronormative societies. This term was coined by Elizabeth Coonrod Martinez, who premised her argument on Sandra Cisneros' 1984 novel *The House on Man*. In describing the condition of multiple oppressions which women live under and endure, Martinez notes:

> women ... do not initiate events in their own lives; instead they endure poverty and racism from the society at large and oppression under the men in their lives. They do not get to choose their spouses, and when they do pick a boyfriend, and get pregnant, they are considered bad girls. They do not have choice – before or after marriage.
>
> (Martinez, 2002: 131)

This aptly captures the many borders which encircle and at times enslave women and from which they can hardly escape. The nature of the patriarchy dictates that they do not initiate any events. In other words, they are loyal followers of their menfolk. Additionally, women and children are the major victims of poverty, famine and drought. These misfortunes render them enclosed within the borders of material deprivation. When it comes to racism as a specific border which confronts women, it must be born in mind that women, especially black women, are the bottom of the ontological ladder.

Probably the greatest gender border is that which prohibits women from choosing their spouses. This is mainly religiously and traditionally prescribed and amounts to the theft of women's agency. Concisely, a gender border is the sum of the negative restrictions placed on women by society; be they structural, institutional, religious, social or economic. Other forms of gender borders exist, such as those explored by Sylvie Vigneux, which consist of the man/woman binary, "not as a pathologized form of 'border crossing,' but instead as simply one facet in an indefinite number of gender variations" (Vigneux, 2011: 13). I will not explore these gender variations in these pages as they have been sufficiently covered in scholarship (Collins, 2009; Feingold, 1994; Martinez, 2002; Morokvasic, 2006; Vigneux, 2011).

The border and coloniality of power

Coined by Anibal Quijano, coloniality of power created the myth of race as an instrument of ordering humanity (into borders) and it espouses the resilient colonial power structures, systems and hegemony of the (former) colonisers which allows them to continue controlling global affairs even with the official end of colonialism. Quijano notes that:

> One of the fundamental axes of this model of power is the social classification of the world's population around the idea of race, a mental construction that expresses the basic experience of colonial domination and pervades the more important dimensions of global power, including its specific rationality: Eurocentrism. The racial axis has a colonial origin and character, but it has proven to be more durable and stable than the colonialism in whose matrix it was established.
>
> (Quijano, 2000: 533)

For those in the (formerly) colonised world, existential options are limited. Either people die from war, famine or other disasters in the zone of non-being or they die while trying to cross the abyssal line into the zone of being. Crossing the abyssal lines assumes many forms. While people may be physically in the zone of being, existentially they remain in the zone of non-being until they "cross the border", that is, become officially recognised through processes such as being given citizenship.

Those migrants on asylum status, while physically in the zone of being, will unfortunately still exist or remain in the zone of non-being (Fanon, 1952; Gordon, 2007; Grosfoguel, Oso and Christou, 2014). While these migrants may have crossed the physical border, there are other covert and overt borders which are more systemic which they have to negotiate (see, e.g., Moyo and Nshimbi, 2017). In this regard, I argue that borders play an ambivalent role, that of including some while excluding others. Borders keep those desired and deemed deserving to be in-country in addition to the citizens while simultaneously excluding those not desired to be in-country.

In order for inclusion and exclusion to be executed, identity plays a central role (Castells, 2004; Cisneros, 2014; Mälksoo, 2016; Ndlovu-Gatsheni, 2009, 2011a). Various markers are used to identify especially the undesired people such as undocumented immigrants. The markers include what is termed "one's status in the republic", religion, country of origin, reason for migration, level of education, skills possessed and potential contribution to the host economy. These qualifiers vary from country to country and even within the same country.

Borders as mutative colonial instruments and the criminalisation of movement

The functions of modern state borders are to authenticate the notion of state sovereignty (Dobler, 2016: 149), while simultaneously maintaining, disciplining and where necessary (re)structuring world order (Chomsky, 1996; Mtapuri et al., 2018; Wallerstein, 2007, 2011). Ordinarily, borders can be argued to be an important instrument of the modern state systems wherein they perform a necessary function in international relations, that of separating countries and other geographic and political administrative regions (Cons and Sanyal, 2013; Dobler, 2016; Lecocq, 2003). It must be admitted that without borders as a formal institution which marks the boundaries between jurisdictions, the international systems could be chaotic (Jones, 2009). The practicality and desirability of a borderless world is one area which needs more analysis and will not be attempted beyond this point.

That wars have been fought by many countries, communities, chieftaincies, clans and families over borders attests to the importance of having borders as official geographical mechanisms for demarcating jurisdictions such as countries, provinces, states and so on (Tilly, 1985: 174). Goran Hyden characterises Africa as a continent with a "dubious reputation of being the world's leading theatre of war, conflict and instability" (Hyden, 2015: 1007). Although the region is not unique, its security threats have become more prevalent. In this way, borders should or are supposed to maintain international order, sustain state sovereignties and allow local and international politics to be predictable and feasible.

It must be noted that life forms have always moved across geographies. By life forms I mean the totality of humans, animals, water, trees and air. These five life forms have for centuries interacted symbiotically until modernity and colonialism, as part of "development" began commodifying and commercialising aspects of these life forms. Let me focus on human movement, which is termed migration in most literature (Cazzato, 2016; Crush, 2000; Johnson, 2017; Kudo, 2015; Stillman et al., 2015), and its concomitant criminalisation, if it is done in ways deemed to be non-conformatory to modern laws and regulations. The criminalisation of human movement is therefore a function of the need to obey abyssal lines.

The Treaty of Westphalia, the Berlin Conference, modernity and colonialism turned borders into institutions complete with supporting mechanisms such as passports, border control agencies, immigration and customs officials and many other related support structures. Such is the need to obey the abyssal line that great efforts are expended to keep those from the zone of non-being away as they try to go "out of bounds", that is, enter into the zone of being "illegally". When (formerly) colonised countries gained political independence, they also inherited the responsibility to manage the abyssal lines on behalf of the (former) colonisers who had by then departed the periphery for the centre, to use Samir Amin's words (Amin, 1976). The challenge in this respect comes when borders are used to maintain the colonial *status quo*. The colonial border as a permanent structure is rationalised and operationalised through various means such as coloniality of power. Those in charge of global affairs (mis)use international laws and institutions and also create new norms and standards in order for them to maintain their colonially gained dominance.

International laws and norms such as the principle of *uti possidetis juris* (as you possess under law) and *terra nullius* (no one's land) were created and used for this purpose. As explained by the International Court of Justice (ICJ), *uti possidetis juris* serves to preserve the boundaries of independent states. As a principle, it is linked to *belli finem* and interpretation leans towards favouring actual possession of land and territory and existing borders and boundaries regardless of how they were acquired. Obviously, this is highly contestable and the merits and demerits of the principle of *uti possidetis juris* will not be analysed beyond this point. Suffice to mention that, there is a single exception in case history where under the auspices of the United Nations, a plebiscite was organised in Nigeria in 1959 for the purpose of determining the status of the English speaking part of Cameroon (Fanso, 1999: 286; Konings and Nyamnjoh, 2018: 31–40). Similar to *uti possidetis juris, terra nullius* was used to occupy what the colonisers deemed to be "no man's land". In this case *man* was defined as European and blacks were not considered human (Benyera et al., 2018; Mignolo, 2009). This principle, like many others, led to the many indigenous peoples in the world losing their land *en masse* and the redrawing of borders without their consent and participation.

Thus, African borders drawn arbitrarily by the colonising European countries, besides being "cast in stone" and reinforced through international principles and norms by those nations with military and economic power, created numerous citizenship and state-building and nation-building challenges for post-independence African states. These include the Great Lakes Region countries of Uganda, Rwanda and Burundi and the Democratic Republic of the Congo (DRC) which have to deal with the Twa who are not officially recognised in any of these states. Similarly, in North Africa, the Berber ethnic Tuareg people of the Sahara are not officially recognised as citizens in Libya,

Algeria, Burkina Faso, Mali or Nigeria. The Twa were originally nomadic and when they settled in various countries were not granted citizenship by any of these countries. In essence, the Twa are landless, stateless people, yet together with the San and the Khoi, they are among the aborigines of Africa (Ndlovu-Gatsheni, 2012: 433; Ndlovu-Gatsheni, 2011b: 12; United States Agency for International Development, 2006). The Tuareg are semi nomadic Berbers and they move across Mali, Libya, Chad, Algeria and Niger (Lecocq, 2003). While mostly concentrated in Mali, they also do not have set citizenship in any of these five countries. For the Twa and the Tuareg, the statement that *borders found us here* and that *borders crossed us* are an apt description of their circumstances (Lamb, 2014; Moyo, 2016; Ndlovu-Gatsheni, 2012: 433).

This should be viewed against the fact that, although it can be argued that conflict in Africa pre-dates both modernity and colonialism, it should be understood that pre-colonial violence in Africa was never about humiliating, exterminating and subjugation. It aimed to achieve the opposite, that is, state-building and nation-building through expanding territories and increasing population and diversity. Regarding the fluidity of pre-colonial and colonial "tribal" borders and the expansion of the Ndebele nation's borders, Julian Cobbing argues:

> The kingdom expanded via the creation of new regiments formed every so often to absorb a new generation of youths, to which were added captives taken in the annual raids upon adjacent Shona tribes. Settlement was thus everywhere.
>
> (Cobbing, 1974: 608)

From the above depiction of the nature of colonial "tribal" borders, there are three characteristics which stand out: borders were fluid, raiding for people and territory was an acceptable form of "international relations", and those defeated became part of the victorious nation wherein they were initiated and accepted as members. With the coming of colonialism, the white form of exterminating violence entered the black body and black on black violence was then termed xenophobia. This form of violence is a colonial creation and is wrongly attributed mainly to black Africans in the zone of non-being (Benyera, 2018). Colonial violence served the purpose of enslaving the conquered, humiliation, segregation and discrimination, that is, what Mahmood Mamdani terms bifurcation of Africans (Mamdani, 2009). In Africa before the arrival of the colonialists, interstate war was never for annihilation and mass murder, rather it was for state expansion where human beings were seen as a resource to be integrated into the winning state. Cases include the Ndebele state which was made up of the Shona and Kalanga peoples who were defeated in war and then assimilated into the Ndebele kingdom.

Indeed, the argument that African borders are needed to exclude those deemed to be undesirables, because people do not move as neutral "vessels", as they move with their cultures, languages, histories and religions, is limited, because it lacks a historical appreciation of African borders before colonial and imperialist conquest. An apt example is the movement of migrants from many countries into South Africa, where they face a "xenophobic" backlash (Benyera, 2018; Rossouw, 2007; Steenkamp, 2009). But, it must be noted that, prior to colonialism and apartheid in South Africa, people moved freely in the sub-region. Mzilikazi left today's KwaZulu Natal, first tried to settle in Thaba Nchu in the Free State, was defeated by a Barolong-Griqua-Afrikaner alliance in 1836 at the battle of Mosega. Mzilikazi tracked north via Mozambique, finally setting his capital in Bulawayo, today's Zimbabwe. While borders were contested, they were not cast in stone and displayed what I can term non-toxic fluidity.

Borders and border thinking

Most of the scholarship on the issue of borders in Africa was concerned with how (post-)colonial Africa must treat the colonially imposed borders (Amadife and Warhola, 1993; Ikome, 2012; Walia, 2013). The great debate on what to do with colonial borders was settled when nationalists won over continentalisms, hence the continued existence of these borders in Africa. The victory by the nationalists over the continentalisms did not resolve the border problem in Africa. Africa's borders have been a source of many conflicts which are well documented. Some are discussed in this volume (see, e.g., Chapter 5). The debate over what to do with colonial borders did not stop human movement especially from the periphery to the centre for various reasons such as fleeing war, conflicts, droughts and economic hardships among other push factors. The result of this human movement which intensified with armed conflicts mainly in North Africa and the Middle East was the deepening of debates between cosmopolitans and nationalists (Cazzato, 2016; Crush, 2000; De Genova, 2016; Mtapuri et al., 2018; Stillman et al., 2015).

How then do we mediate this conflict where cosmopolitanism clashes with nationalism? What I am advancing here is not new to scholarship but has been discussed by the likes of Walter D. Mignolo and is termed border gnosis (Mignolo, 2000). Practically, how does border thinking work, especially regarding the issue of borders, migrants and human movement? Border thinking is operationalised through seeing oneself from the position of the other. That is, seeing the world from the perspective of the other and seeing the other as the other sees themselves. This translates to true and authentic humanity, humanism and humanness. Put simply, border gnosis or border thinking involves the bringing together of different cultures without causing conflict. Conflict is avoided by accepting differences and in the process coexisting without imposition of one culture over the other. Geographical spaces where cultures and languages learn from one another and thrive have always existed and tended to find mutual modes of coexistence. The people of the

Amazon rainforest in South America have always moved across geographies with little or no major contestations. A lot therefore can be learnt about borders and the management of human movement from these peoples. Stated differently, what is being advocated in these pages is the notion of a pluriversality.

Conclusion

This edited volume is aptly named, *African Borders, Conflict, Regional and Continental Integration*. The idea behind its conceptualisation is to recognise that while African borders are indeed a colonial legacy which has contributed to wars and conflicts in Africa, there are possibilities of using Africa's cross-border architecture and dynamics for peace. The aim of the book is to reign-ite the reimagination of a new Africa where human movement is not an offence but a resource. The overarching theme of the book is therefore to encourage a thinking of borders in Africa from an African locus of enuncia-tion. Throughout the book, borders, wars and conflicts in Africa are pre-sented and analysed from the viewpoint of those whose lived experiences suggest that the African borders could be a space where peace and develop-ment spring from and not war.

Currently, there are mixed attitudes towards continentalism and isolation with the USA under President Donald Trump, moving more towards isolation from the international system of nations. In mainland Europe, continentalism is on the retreat and yet the same cannot be said about Africa where regional integration and continentalism are in vogue. The transformation of the Organisation of African Unity (OAU) to the African Union (AU) indicates that Africa's moves towards regional integration are bearing fruit, as wit-nessed by the end of the conflict between Eritrea and Ethiopia. To counter the divisive effects of borders, African countries are currently engaged in the regional integration project, through Regional Economic Communities (RECs). Regional integration and continentalism are efforts meant to create larger geographical administrative areas and in a way reduce the role of national borders as goods, services and people would be free to move across these borders. This was the idea when the OAU was formed. The same idea was renewed when the AU took over the mandate of the OAU. These insti-tutional arrangements promise to reduce the impact of or "redraw" African borders both as a mechanism of forging a Pan African identity and also unleashing the development potential of the continent. The narrative of Africa as a war torn continent is therefore a deliberate exaggeration of the reality as there are major strides towards sustainable peace on the continent which are documented in this book. While Africa is yet to attain sustainable peace, presenting the continent as a war torn place negates the many success strides such as the way in which the Gambian disputed election results of 2017 was handled by the regional body, the Economic Community of West African States (ECOWAS). In this edited volume, the authors respond to

Eurocentric interpretations of African wars and conflicts by offering a decolonial response to the debate.

In this chapter, I have argued that borders have various uses and interpretations depending on where the analyst is located both geographically and epistemologically. My main argument is that borders are not innocent infrastructure which merely indicates where one country ends and where another country begins. Borders in Africa, as products of the Berlin Conference, continue to play the role of maintaining a bifurcated humanity reinforced by coloniality of power where one set of people exists in the zone of being and the others in the zone of non-being. Those in the zone of non-being experience the darker side of modernity, slavery and colonialism and are now the subject of many debates as they continue perishing in the Mediterranean Sea in their attempt to migrate to Europe. This has resulted in borders assuming a central position. Using the locus of enunciation, it was argued that those analysing borders from a Western-centric perspective view them as protective instruments which keep undesired human elements out of their countries. Those from the (formerly) colonised world view borders, especially borders located in Africa, as institutions meant to reinforce and perpetuate their subjugation.

References

Amadife, Emmanuel N. and Warhola, James W. 1993. Africa's Political Boundaries: Colonial Cartography, the OAU, and the Advisability of Ethno-National Adjustment. *International Journal of Politics, Culture, and Society* 6(4): 533–554.

Amin, Samir. 1976. *Unequal Development: An Essay on the Social Formations of Peripheral Capitalism*. New York: Monthly Review Press.

Anon. n.d. Predicting the Brexit Vote: Getting the Geography Right (More or Less). http://blogs.lse.ac.uk/politicsandpolicy/the-brexit-vote-getting-the-geography-more-or-less-right/.

Benyera, Everisto. 2017. Towards an Explanation of the Recurrence of Military Coups in Lesotho. *Air & Space Power Journal – Africa and Franchophonie* 8(3): 56–73.

Benyera, Everisto. 2018. The Xenophobia-Coloniality Nexus: Zimbabwe's Experience. In A. O. Akinola (Ed.), *The Political Economy of Xenophobia in Africa*. Cham, Switzerland: Springer International Publishing, pp. 135–151.

Benyera, Everisto, Mtapuri, Oliver and Nhemachena, Artwell. 2018. The Man, Human Rights, Transitional Justice and African Jurisprudence in the Twenty-First Century. In

Castells, Manuel. 2004. *The Power of Identity*. Malden, MO: Blackwell Publishers.

Cazzato, Luigi. 2016. Mediterranean: Coloniality, Migration and Decolonial Practices. *Politics. Rivista Di Studi Politici* 1(5): 1–17.

Chinweizu, Ibekwe. 1993. Reparations and a New Global Order: A Comparative Overview. Abuja: Pan-African Conference on Reparations.

Chomsky, Noam. 1996. *World Orders, Old and New*. New York: Columbia Univerity Press.

Cisneros, Josue David. 2014. *The Border Crossed Us: Rhetorics of Borders, Citizenship, and Latina/o Identity*. Tuscaloosa: University of Alabama Press.

Clark, Phil. 2007. Hybridity, Holism, and "Traditional" Justice: The Case of the Gacaca Courts in Post-Genocide Rwanda. *George Washington International Law Review* 1(39): 765–837.

Cobbing, Julian. 1974. The Evolution of Ndebele Amabutho. *Journal of African History* 15(4): 607–631.

Collins, Kimberly. 2009. Introduction to "Border and Gender Studies: Theoretical and Empirical Overlap". *Eurasia Border Review: Border and Gender Studies* 7(1): 51–53.

Cons, J. and Sanyal, R. 2013. Geographies at the Margins: Borders in South Asia – an Introduction. *Political Geography* 35: 5–13.

Crush, Jonathan. 2000. The Dark Side of Democracy: Migration, Xenophobia and Human Rights in South Africa. *International Migration* 38(6): 89–91.

Curtice, John. 2016. Brexit: Behind the Referendum. *Political Insight* 7(2): 4–7.

De Genova, Nicholas. 2016. The "European" Question: Migration, Race, and Post-Coloniality in "Europe". In Anna Amelina, Kenneth Horvath and Bruno Meeus (Eds), *An Anthology of Migration and Social Transformation*. Cham, Switzerland: Springer, pp. 343–356.

Dobler, Gregor. 2016. The Green, the Grey and the Blue: A Typology of Cross-Border Trade in Africa. *Journal of Modern African Studies* 54(1): 145–169.

Fanon, Frantz. 1952. *Black Skin White Masks*. London: Pluto Press.

Fanso, Verkijika G. 1999. Anglophone and Francophone Nationalisms in Cameroon. *The Round Table* 88(350): 281–296.

Feingold, Alan. 1994. Gender Differences in Personality: A Meta-Analysis. *Psychological Bulletin* 116(3): 429–456.

Gordon, Lewis R. 2007. Through the Hellish Zone of Nonbeing: Thinking through Fanon, Disaster, and the Damned of the Earth. *Human Architecture: Journal of the Sociology of Self-Knowledge* 5(Summer): 5–12.

Grosfoguel, Ramón. 2011. Decolonizing Post-Colonial Studies and Paradigms of Political-Economy: Transmodernity, Decolonial Thinking, and Global Coloniality. *TRANSMODERNITY: Journal of Peripheral Cultural Production of the Luso-Hispanic World* 1(1): 1–38.

Grosfoguel, Ramón. 2015. What Is Racism? *Journal of World-Systems Research* 22(1): 9–15.

Grosfoguel, Ramón and Cervantes-Rodriguez, Ana Margarita (Eds). 2002. *The Modern/Colonial/Capitalist World-System in the Twentieth Century: Global Processes, Antisytemic Movements, and the Geopolitics of Knowledge*. Westport, CT: Greenwood Press.

Grosfoguel, Ramón, Oso, Laura and Christou, Anastasia. 2014. "Racism", Intersectionality and Migration Studies: Framing Some Theoretical Reflections. *Identities* 22(6): 1–18.

Hobolt, Sara B. 2016. The Brexit Vote: A Divided Nation, a Divided Continent. *Journal of European Public Policy* 23(9): 1259–1277.

Hyden, Goran. 2015. Rethinking Justice and Institutions in African Peacebuilding. *Third World Quarterly* 36(5): 1007–1022.

Ikome, Francis Nguendi. 2012. Africa's International Borders as Potential Sources of Conflict and Future Threats to Peace and Security. Institute for Security Studies Paper 233: 1–16.

Johnson, Jessica A. 2017. After the Mines: The Changing Social and Economic Landscape of Malawi–South Africa Migration. *Review of African Political Economy* 44(152): 237–251.

Jones, R. 2009. Categories, Borders and Boundaries. *Progress in Human Geography* 33: 174–189.

Konings, Piet and Nyamnjoh, Francis B. 2018. *Negotiating an Anglophone Identity: A Study of the Politics of Recognition and Representation in Cameroon.* Leiden: Afrika-Studiecentrum Series.

Kudo, Yuya. 2015. Female Migration for Marriage: Implications from the Land Reform in Rural Tanzania. *World Development* 65: 41–61.

Laine, J. 2015. A Historical View on the Study of Borders. In S. Sevastianov, J. Laine and A. Kireev (Eds). *Introduction to Border Studies.* Vladivostok: Far Eastern Federal University, pp. 15–32

Lamb, V. 2014. "Where Is the Border?" Villagers, Environmental Consultants and the "Work" of the Thai–Burma Border. *Political Geography* 40: 1–12.

Larner, Wendy. 2015. Globalising Knowledge Networks: Universities, Diaspora Strategies, and Academic Intermediaries. *Geoforum* 59: 197–205.

Lazer, David. 2011. Networks in Political Science: Back to the Future. *PS: Political Science & Politics* 44(1): 61–68.

Lecocq, Baz. 2003. This Country Is Your Country: Territory, Borders, and Decentralisation in Tuareg Politics. *Itinerario* 27(1): 59–78.

Lester, Alan. 2001. *Imperial Networks: Creating Identities in Nineteenth-Century South Africa and Britain.* London: Routledge.

Maldonado-Torres, Nelson. 2016. Outline of Ten Theses on Coloniality and Decoloniality. *Berkeley Planning Journal* 26(1): 1–37.

Mälksoo, Maria. 2016. State Identity Disjuncture and the Politics of Transitional Justice: The Case of Russia. CEEISA-ISA Joint Conference, Ljubljana.

Mamdani, Mahmood. 2009. *Saviors and Survivors: Darfur, Politics and the War on Terror.* London: Verso.

Mamdani, Mahmood. 2010. Responsibility to Protect or Right to Punish? *Journal of Intervention and Statebuilding* 4(1): 53–67.

Mamdani, Mahmood. 2015. Political Identity, Citizenship and Ethnicity and Post-Colonial Africa. Keynote address at the Conference "New Frontiers of Social Policy". Arusha.

Martinez, Elizabeth Coonrod. 2002. Crossing Gender Borders: Sexual Relations and Chicana Artistic Identity. *Melus* 27(1): 131–148.

Mazrui, Ali A. 1986. *The African: A Triple Heritage.* London: British Broadcasting Corporation.

Mazrui, Ali A. 1994. Comment: Africa: In Search of Self-Pacification. *African Affairs* 93(370): 39–42.

Mazrui, Ali A. 2005. Pan-Africanism and the Intellectuals: Rise, Decline and Revival. In T. Mkandawire (Ed.), *African Intellectuals: Rethinking Politics, Language, Gender and Development.* Dakar, Senegal: Council for the Development of Social Science Research in Africa (CODESRIA) and Zed Books, pp. 56–77.

Mbembe, Achille. 2000. *On Private Indirect Government: State of the Literature Series Number 1.* Dakar, Senegal: CODESRIA Books.

McEwen, Nicola. 2016. Disunited Kingdom: Will Brexit Spark the Disintegration of the UK? *Political Insight* 7(2): 22–23.

Mignolo, Walter D. 2000. *Local Histories/Global Designs: Coloniality, Subaltern Knowledges, and Border Thinking.* Princeton, NJ: Princeton University Press.

Mignolo, Walter. 2009. Who Speaks for the "Human" in Human Rights? *Hispanic Issues on Line* 5(1): 7–24.

Moyo, Inocent. 2016. The Beitbridge–Mussina Interface: Towards Flexible Citizenship, Sovereignty and Territoriality at the Border. *Journal of Borderlands Studies* 31(4): 427–440.

Moyo, Inocent and Nshimbi, Christopher Changwe. 2017. Of Borders and Fortresses: Attitudes Towards Immigrants from the SADC Region in South Africa as a Critical Factor in the Integration of Southern Africa. *Journal of Borderlands Studies*. DOI: doi:10.1080/08865655.2017.1402198.

Morokvasic, Mirjana. 2006. Crossing Borders and Shifting Boundaries of Belonging in Post-Wall Europe. A Gender Lens. *Grenzüberschreitungen—Grenzziehungen: Implikationen Für Innovation Und Identität*, 47–72. Retrieved from http://aa.ecn.cz/img_upload/6334c0c7298d6b396d213ccd19be5999/MMorokvasic_Crossingbordersandshiftingboundaries.pdf.

Mpofu, W. 2014. A Decolonial "African Mode of Self-Writing": The Case of Chinua Achebe in Things Fall Apart. *New Contree* 69(July): 1–25.

Mtapuri, Oliver, Nhemachena, Artwell and Benyera, Everisto. 2018. Towards a Jurisprudential Theory of Migration, Foot-Looseness and Nimble-Footedness: The New World Order or Pan-Africanism? In A. Nhemachena, T. V. Warikandwa and S. K. Amoo (Eds), *Social and Legal Theory in the Age of Decoloniality: (Re-)Envisioning African Jurisprudence in the 21st Century*. Bamenda, Cameroon: Langaa, pp. 236–298.

Ndlovu-Gatsheni, Sabelo J. 2009. *Do "Zimbabweans" Exist? Trajectories of Nationalism, National Identity Formation and Crisis in a Postcolonial State*. Bern: Peter Lang.

Ndlovu-Gatsheni, Sabelo J. 2010. Do "Africans" Exist? Genealogies and Paradoxes of African Identities and the Discourses of Nativism and Xenophobia. *African Identities* 8(March): 281–295.

Ndlovu-Gatsheni, Sabelo J. 2011a. African, Know Thyself: Epistemic Awakening and the Creation of an Identity for the Pan African University. Plenary Presentation delivered at the Pan African University Curriculum Review Workshop organised by the African Union Commission, 14–18 November, Addis Ababa.

Ndlovu-Gatsheni, Sabelo J. 2011b. The Logic of Violence in Africa. Ferguson Centre for African and Asian Studies Working Paper No. 2. PDF available online.

Ndlovu-Gatsheni, Sabelo J. 2012. Beyond the Equator There are No Sins: Coloniality and Violence in Africa. *Journal of Developing Societies* 28(4): 419–440.

Ndlovu-Gatsheni, Sabelo J. 2013. *Coloniality of Power in Postcolonial Africa: Myths of Decolonization*. Dakar, Senegal: CODESRIA Books.

Ndlovu-Gatsheni, Sabelo J. 2015. Decoloniality as the Future of Africa. *History Compass* 13(10): 485–496.

Nhemachena, Artwell, Warikandwa, Tapiwa Victor and Amoo, Samuel K. (Eds), *Social and Legal Theory in the Age of Decoloniality: (Re-)Envisioning African Jurisprudence in the 21st Century*. Bamenda, Cameroon: Langaa, pp. 187–218.

Nimako, Kwame. 2015. Reorienting the World: With or Without Africa. *Journal of World-Systems Research* 21(ReOrienting the World: Decolonial Horizons): 193–202.

Novak, P. 2011. The Flexible Territoriality of Borders. *Geopolitics* 16: 741–767.

Paasi, A. 2012. Border Studies Reanimated: Going beyond the Territorial–Relational Divide. *Environment and Planning* 44: 2303–2309.

Quijano, Anibal. 2000. Coloniality of Power, Eurocentrism, and Latin America. *Nepantla: Views from South* 1(3): 533–580.

Ranger, Terence. 1989. Missionaries, Migrants and the Manyika: The Invention of Ethnicity in Zimbabwe". In L. Vail (Ed.), *The Creation of Tribalism in Southern Africa*. London / Berkeley: Currey / University of California Press, pp. 118–151.

Rossouw, Gideon. 2007. Xenophobia at the End of the Rainbow: Explaining Anti-Foreigner Violence in the Context of Limits to the South African Rechtsstaat. PDF available online.

Senier, Amy. 2008. Traditional Justice as Transitional Justice: A Comparative Case Study of Rwanda and East Timor. *PRAXIS The Fletcher Journal of Human Security* 13: 67–88.

de Sousa Santos, Boaventura. 2007. Beyond Abyssal Thinking: From Global Lines to Ecologies of Knowledges. *Review (Fernand Braudel Center)* 30(1): 45–89.

Statistics South Africa. 2015. *Vulnerable Groups Indicator Report 2016*. Pretoria: Statistics South Africa.

Steenkamp, Christina. 2009. Xenophobia in South Africa: What Does It Say about Trust? *The Round Table* 98(403): 439–447.

Stillman, Steven, Gibson, John, McKenzie, David and Rohorua, Halahingano. 2015. Miserable Migrants? Natural Experiment Evidence on International Migration and Objective and Subjective Well-Being. *World Development* 65: 79–93.

Tafira, Chimusoro Kenneth and Ndlovu-Gatsheni, Sabelo J. 2017. Beyond Coloniality of Markets – Exploring the Neglected Dimensions of the Land Question from Endogenous African Decolonial Epistemological Perspectives. *Africa Insight* 46(4): 9–24.

Táíwò, Olúfémi. 2009. *How Colonialism Preempted Modernity in Africa*. Bloomington: Indiana University Press.

Tilly, Charles. 1985. War Making and State Making as Organized Crime. In Peter Evans, Dietrich Rueschemeyer and Theda Skocpol (Eds), *Bringing The State Back In*. Cambridge: Cambridge University Press, pp. 169–187.

United States Agency for International Development. 2006. Reconstruction for Development in Burundi: Guiding Criteria and Selected Key Issues. (May): 1–30.

Vigneux, Sylvie. 2011. Rethinking Gender Borders: The Role of Puberty in Creating a Universalist Model of Gender Identity. *Footnotes* 4: 10–14.

Walia, Harsha. 2013. *Undoing Border Imperialism*. Oakland, CA: AK Press.

Wallerstein, Immanuel. 2007. *World Systems Analysis: An Introduction*. Durham, NC: Duke University Press.

Wallerstein, Immanuel. 2011. *The Modern World-System: Capitalist Agriculture and the Origins of the European World-Economy in the Sixteenth Century, Volume 1*. Berkeley: University of California Press.

Weiss, Linda. 1998. *The Myth of the Powerless State. Governing the Economy in the Global Era*. Cambridge: Polity Press.

Weiss, Linda. 2000. Globalization and State Power. *Development and Society* 29(1): 1–15.

Wiegink, Nikkie. 2015. Former Military Networks a Threat to Peace? The Demobilisation and Remobilisation of Renamo in Central Mozambique. *Stability: International Journal of Security & Development* 4(1): 1–16.

Willsher, Kim. 2018. "Spider-Man" of Paris to Get French Citizenship after Child Rescue: President Macron Thanks Malian Migrant Who Climbed Four Storeys to Save Boy. *The Guardian*, 28 May.

3 Trans-border trajectories of violence

Capabilities for peace and cooperative engagement in the Afar Horn

Muauz Gidey Alemu

Introduction

Boundaries, borders and borderlands in the Afar Horn are sources of tension, belligerence and insecurity among Afar Horn states: Ethiopia, Eritrea and Djibouti. Borders create different constraints and many academics have argued about this from different vantage points (Dereje and Hoehne, 2008: 7). This chapter, however, focuses on an analytical framework dealing with the opportunity side of the border "coin". The central contention of the study is to show how failure on the part of the centre to avail itself of the opportunities provided by the peripheries for peace has transformed the unutilised opportunities into trajectories of violence. The focus of this chapter is neither what state borders have done to the people, nor how borderlanders make use of them as conduits and opportunities, but the failure on the part of states to recognise and avail themselves of the opportunities and capabilities for cooperation which turn these trajectories of cooperation into trajectories of violence.

The first part defines the concepts of boundary, border, borderlands, resourcing borders, violence and peacebuilding. The second part describes the nature of borders and borderlands in the Afar Horn in comparison with the peculiarity of borders and borderlands in the Horn of Africa. The third part examines the various opportunities for peace and cooperation provided by borders and borderlands informally utilised by the borderlanders but overlooked by the Afar Horn states. The fourth part analyses how the same unused opportunities are changed into trajectories of violence and insecurity in inter-state relations. The fifth part reflects on rethinking our understanding of borders and borderlands as capabilities for regional peace.

Conceptual framework

The concepts of boundary, border and borderland, as well as resourcing borders used here, require a brief description. The definitions carry a predisposition towards supporting a rigid conception of the role and functions of boundary, border and borderland. However, these concepts are used here, cognisant of their positive and negative roles and the potential for using them for constructive ends as presented by the concept of resourcing borders.

Boundary, border and borderland

Boundary refers to the distinctive line delineating where the sovereignty of one state ends and that of another begins. This is the line commonly known as an "international boundary", and is often demarcated by physical marks and limits. In the Westphalian state system, a clearly defined boundary constituted the hallmark of the independence, sovereignty and legitimacy of a state to the international community of states. According to Wafula Okumu, a boundary is the territorial and physical limit of a state's jurisdiction. A boundary (like the one found among the Afar people) divides contiguous territories and along with it, separates one identity group into different state arrangements. There are two types of boundary: one is fixed, which is precisely demarcated and can be re-demarcated in the same way in the event that beacons of boundary are lost. The second, general boundary is an undetermined line between two states. Boundaries are determined by mutually agreed international treaties among states which thereafter serve as physical marks of controlling the movement of people and goods entering a state's territory (Okumu, n.d.: 1–2). The nature of boundaries in the Afar Horn are both fixed in the sense that they have exact lines but also general, because the legitimacy of the boundaries is contested.

Border refers to a region straddling a boundary which does not necessarily have to be exactly on the boundary. It is also defined as any place with the range of 12 nautical miles of a state's legal jurisdiction. Therefore, it determines citizenship, rights and privileges, way of life and destiny of a people inside it as different from those outside it. Borderland is a related term which refers to the borders (zones adjacent to a boundary) the life of whose inhabitants are influenced by their interaction with their neighbours across the border or on the other side of the boundary (ibid.). According to John Agnew, for various purposes and utilities, borders may involve:

> the fencing of a chunk of territory separating people. Borders may also be artefacts of dominant discursive processes. The discursive utility of borders may change and as they do, borders live on as residual phenomena that may still capture the imagination but no longer serve essential purposes.
>
> (Agnew, 2008: 175)

This underscores the construction, deconstruction and reconstruction of borders' political discourse, decisions or cultural and artistic representations (Durand, 2015: 2). So, this conception underscores that borders mimic human creation, which may have both instrumental and non-instrumental utilities. For the purpose of this study, borders are contested boundaries but inhabited by people sharing one identity and a contiguous territory. Therefore, border is an appropriate term in place of boundary. The remarkable interdependence and active interaction among the Afar people in the region make them a community of borderlanders. This calls for defining a further concept: resourcing borders.

Resourcing borders

Resourcing borders is the utilisation of the capabilities and inhibitions created by the nature of borders and borderlands by the borderlanders. This recognises that borders are zones of dynamic interaction for the construction, maintenance and consolidation of which, multiple actors are involved. Activities across borders like smuggling, for Paul Nugent, neither involve resistance, nor the continuation of old trade relations, but rather a natural interaction:

> The very creation of the boundary was bound to have an impact on the local economic geography, opening up avenues of profitable commerce where they had not previously existed. Apart from opening up new trade routes, the smuggling complex also summons forth a new breed of entrepreneurs whose very livelihoods depended upon the perpetuation of the international boundary.
>
> (Nugent, 2002: 12)

New opportunities follow old trade routes and social infrastructure in the Afar Horn. Moreover, the social space across borders (not fully owing to the artificiality of colonial boundaries or the inconsequential nature of boundaries to the borderlanders) facilitates smooth interaction in line with Nugent's argument. This is best represented through an understanding of arbitrage economies which are cross-border economic activities created and sustained by the presence of state borders. This considers the existence of divergent economic and administrative regimes, price and market conditions that provide an opportunity for people residing across the border (Anderson and O'Dowd, 1999: 681).

However, this is not always true: resourcing borders, besides being a natural process emanating from the existence of borders or opportunities created by borders, can be a mode of political resistance. In the context of marginalisation of, and resistance by, borderlanders against the centre, cross-border activities like smuggling, contraband and other types of resourcing borders can be forms of political resistance. Resourcing borders could be a means of political resistance to the centre which is a continuation of a centre–periphery divide in the Afar Horn. A more comprehensive analytical framework for resourcing borders is provided by the work of Dereje and Hoehne based on the analysis of resourcing borders in the Horn of Africa. They write:

We identify four different types of resources that collective as well as individual actors can extract from state borders and borderlands. These are, first, economic resources (cross-border trade and smuggling); second, political resources (access to alternative centres of political power; trans-border political mobilization; sanctuary for rebels who strive to alter national structures of power; and strategic co-option of borderlanders by competing states); third, identity resources (state border as security device in an inter-ethnic competition; legitimation of the claim for statehood); and fourth, status and rights resources (citizenship and refugee status, including access to social services such as education).

(Dereje and Hoehne, 2008: 9)

The four resources identified by Dereje and Hoehne provide an apt framework for analysing resourcing borders in the Afar Horn. The opportunities provided by borders and borderlands to borderlanders are the result of two major factors. The first is the inter-state rigidity of what borders represent. States conceive borders as given and fixed facts of life and attempt to sustain their status. The second is the fluidity and permeability of borders for borderlanders. In this sense, the state borders are simultaneously permeable and consequential. Dereje and Hoehne (2008) recognise that the relative position of borderlanders in the national power calculus and the perception by the state of borderlanders, form vital analytical factors to be considered in examining the resourcing of borders.

The contradictory conception of resourcing borders for states (as violations of sovereignty) and borderlanders (as legitimate opportunities created by the coming into existence of borders) is rooted in the polarised conceptions that consider borders as enabling, and disabling, structures. This is well elaborated by John Agnew. Competing views of borders are based on normative assumptions as opposed to empirical orientations. One side looks at a border as fixed, and *a priori* from an ontological point of view. This view conceives borders as lasting lines or structures meant for a certain purpose. Thus, it prescribes thinking about borders from the point of view of the political, economic and social purposes they serve and how they serve them. The second perspective considers borders as artefacts of a dominant discursive process which undergoes processes of change (not to be considered as given) with no essential purpose. This view considers borders as the product of human creation which should at any time be open to scrutiny. Borders, viewed in the most extreme cases, can be the product of historical processes anachronistic to the present, or processes rendered invisible by transnational, and global, pressures. Therefore, borders devised through human action are made to be structures which construct thoughts of alternative political, economic, and social possibilities (Agnew, 2008). He argues that:

thinking about borders should be opened up to considering borders as "dwelling" rather than as national space and political responsibility for pursuit of a "decent" life as extending beyond the borders of a particular state. Borders matter both because they have real effects and because they trap thinking about and acting in the world in territorial terms.

(Agnew, 2008: 175)

Thinking about borders should be released from rigid frameworks to reimagine multiple alternative ways of thinking, creating, recreating and utilising them. This means reimagining borders to enable the enrichment of human life. The conception applied here is neither the totally *a priori* and ontological conception nor the one which views borders as artefacts of discursive processes (in constant flux) which must be overcome. They are conceived rather as zones of multiple possibilities and capabilities for human realisation according to which, they must be imagined, evaluated, acted upon and used. Concurrent to Agnew's ethical commitment, the normative commitment of this study is that, borders should be thought of and utilised in the promotion of human and state security as capabilities for cross-border peace, shared by peoples and states, as opposed to trajectories of cross-border violence. In Jonathan Seglow's wording, in the promotion of the "right to a decent life" (Agnew, 2008: 176), in the context of borders in the Afar Horn running between the Afar people who share the same view and purpose of border (real and discursive utilities), the people deserved to reap the dividends of inter-state peace. This requires transcending the binary conceptual divide and transcending the current limitations on intellectual, imaginative and political capabilities to find a more practical way of imagining borders as zones of capability. Therefore, this chapter applies Agnew's conceptual framework of borders – which views borders as zones of changes and transformations, and spaces of multiple possibilities and capabilities for human realisation – as its major conceptual framework.

Related to the above conception of borders and the nature of Afar Horn borders in particular, is the idea of *production of cross-border space*. According to Frédéric Durand, production of cross-border space is the restructuring and projecting of borders and borderlands towards the outside world to create an institutionalised partnership for the common good of the people across the borders (Durand, 2015: 3). The production of cross-border space presumes dynamic borders capable of responding to basic human needs and acts as capabilities for cooperation. The Afar Horn is an authentic cross-border space built by multiple factors. The nature and utility of this space (constituting trajectories of violence and capabilities for peace) is investigated from the vantage point of decreasing violence and promoting peace.

Violence and peacebuilding

Violence is any form of physical and somatic harm which inhibits the realisation of human potential. This includes direct war violence, structural violence (among others in the form of marginalisation, oppression and destitution) and cultural violence (which provides justification for war and structural violence). In all forms, Johan Galtung considers violence as an insult to basic human needs which includes various types of insecurity (Galtung, 1996: 169; Galtung, 1990: 294). The trajectories of violence, accordingly, refers to the various activities such as inter-state war, insurgency, contraband, human and arms trafficking, and political marginalisation and economic destitution of the people in the Afar Horn. As borders should be infrastructures of pursuing a decent life, the absence of trajectories of violence and their replacement by capabilities for peace is a mandatory task of peacebuilding.

Peace is the absence of violence in all forms. Positively speaking, peace is a condition of harmony, cooperative engagement, non-violence and the presence of enabling conditions for the realisation of human potential (Galtung, 1969: 170). According to Galtung and John Lederach, peacebuilding is a long time process of moving away from violence to harmonious relations through constructive engagement. This involves the process of identifying and utilising the capabilities for peace and transforming structures of violence to structures of cooperative engagement (Galtung, 2000; Lederach, 1999). Galtung and Lederach use different terms to represent similar sets of ideas about creative engagement regarding the violent condition and the ability to envision as well as create a better reality out of the experience of violence. This process is recognised as transcendence for Galtung, and transformation for Lederach. However, both conceive the idea of creatively imagining and envisioning beyond the immediacy of the danger of episodes of violence towards a better future. It is about addressing both the immediate basic human needs, and strategically thinking about long term changes. It underscores both changing the quality of the relationship, structures and value systems supporting violence, and thinking beyond the immediate conditions of violence (see also Lederach, 2003, 2005). In the context of borders, this means the dual task of transformation of cross-border structures of violence to infrastructures for peace and the utilisation of cross-border capabilities for peace to change the prevailing violent relations.

According to Galtung, "violent structures leave their mark not only on the human body but also on the mind and spirit of human beings". Such structures are effective tools for crippling the anti-exploitation struggles of the oppressed, and impeding consciousness formation and mobilisation which have to be overcome to build peace (Galtung, 1990: 294). Lederach similarly emphasises the quality of relationships in society, how close and far people wish to be together, as an important point towards reducing violence and increasing justice (Lederach, 1999: 12). The presence of structures obstructing normal relationships in society includes the potential for direct violence against individual victims, which has to be creatively engaged to build peace. Therefore, borders can be viewed as cross-border spaces composed of both violent structures, and peace capabilities, which should be constructively engaged for peacebuilding. Lederach elaborated on peacebuilding in terms of three interrelated strategies. One is to transcend destructively violent patterns of relations while still living in the context that produced them. The second is to build broad processes of social change while creating genuine spaces of accessible public engagement in dialogue. The third is to promote structural change that can be translated into visible constructive and cooperative action (Lederach, 2005: 61). Therefore, Afar Horn borders are to be imagined and envisioned as genuine spaces and authentic zones for peacebuilding.

This requires dialogical (shared) envisioning of the best possible scenarios of the future, identifying the capabilities for realising the envisioned future, and engaging in cooperative dialogue and action towards changing the quality of relations. It also requires recognising authentic capabilities, opening authentic dialogue and institutionalising new processes and ways of imagining the future accepted by the people involved, by going beyond current challenges and constraints (ibid.). Galtung and Lederach identify the need for creating spaces for engaging in peacebuilding as vital for the transformation of relations and looking at current predicaments as opportunities for constructive transformation of relations. Thus, the peacebuilding framework is used in analysing and reimagining Afar Horn borders and borderlands as capabilities for peace.

The nature of boundaries, borders and borderlands in the Afar Horn

The Horn of African has been one of the regions experiencing the most complex intra- and inter-state conflicts involving boundary issues in the post-independence period. The strategy and politics of mutual insurgency and destabilisation common in the Horn of Africa has been facilitated by the porous nature of the borders that allows easy flow of insurgents, weapons and refugees (Lyons, 1996: 85–87; Cliffe, 1999: 89). According to an Institute of Security Studies' (ISS) report, out of 22 inter-state conflicts involving boundaries from 1950 to 2000, seven occurred in the Horn of Africa. Six out of the seven conflicts have involved the two Afar Horn states of Ethiopia and Eritrea. The percentage share of conflicts involving borders and borderlands of the Horn region in general is 32 per cent, of which the Afar Horn constitutes more than 27 per cent (Ikome, 2012: 4).

The overall picture of the nature of borders and borderlands depicted above, as much as it is true, is overburdened with exaggerated pessimism towards the possibility of managing and utilising borders and borderlands. This is more so, given the fragility of international boundaries and the conflicts arising therefrom in the Afar Horn sub-region. The boundaries between the three states cause various degrees of difficulty with respect to inter-state relations, for the communities of the affected areas. The overall characterisation of post-colonial African state boundaries as artificial and porous may not fully explain the problems of borders and borderlands in this part of Africa. Also important are the complex interplay of factors arising from the nature of local politics and the respective foreign relations of the states, as much as the lack of amicable relations between them. As good borders are those drawn between good neighbours, the lack of management and utilisation of boundaries has made them more of a liability than an asset. Therefore, the major explanation for the problem of borders is associated with the failure of the Afar Horn states to utilise cross-border capabilities for peace in the pursuit of a decent life for borderlanders on both sides of the borders.

Partly, the problem is rooted in the continental debate about the legitimacy of boundaries of post-colonial African states – the reversibility versus the irreversibility of boundaries (ibid.: 3). This polarised approach has been reflected in the policy of Afar Horn states. Eritrea follows the reversibility of boundaries it shares with Ethiopian and Eritrea, while Djibouti and Ethiopia follow the "sacrosanct" nature of colonial borders stipulated by the Organisation of African Unity (OAU) Cairo declaration. Their policies, however, are not monolithic in the sense that all states do not pursue similar policies of either one or the other. Eritrea, the youngest state in the Afar Horn, follows a rigid revisionist policy of re-bordering with all its neighbours (Medhane, 2008). Eritrean policy of redrawing borders is promoted by its belligerent stance and militaristic approach against Ethiopia during the Ethio-Eritrea Border War (EEBW, in 1998), Djibouti during the Ras Doumeirah (1996 and 2008), Yemen (1995) and the Sudan (1994) (Berouk, 2008: 4).

The issue of borders and borderlands in Eritrea is part of the overall Eritrean nation building project. The task of building a unified Eritrean identity out of the country's nine ethnic groups was based on redrawing its international borders by blood shedding. This is a continuation of the Eritrean People's Liberation Front's (EPLF) policy followed during the armed struggle and post-independence. The EPLF, an armed movement, was committed to the birth of the Eritrean state and hence the redefinition of its border with Ethiopia, demonstrating its rejection of the 1964 Cairo Declaration of keeping colonial boundaries intact and sacrosanct (OAU, 1964). As though to settle historical scores against the Cairo Declaration made 30 years before, President Issayas of Eritrea in his inaugural speech in 1993 at an OAU summit in Cairo, denigrated the OAU for failing to support Eritrea's 30-year struggle for independence (Berouk, 2008: 4). This was an unequivocal rejection of the Cairo Declaration engineered to rescue post-colonial states from being bogged down in endless boundary conflicts. As part of its nation- and state-building project, Eritrea pursued the policy of redefining borders, and the shared identity of borderlanders (Medhane, 1999). Consequently, it propagated the politics of Eritrean exceptionalism, characterised by jingoistic state nationalism. It waged border wars with all its neighbours except Saudi Arabia.

The Afar homeland shared by the three states is characterised by contiguous territory settled by the Afar people who, possessing a unified identity, seldom recognise the validity of state borders. They consider state borders running among the three, closely knit Afar clans, as a mark of their predicament. This accounts for their domination by the Issa in Djibouti, their vulnerability to various Ethio-national border conflicts in Ethiopia, and their repression under the Eritrean regime. Despite the divergent policies of the three states and the condition of the Afar people in the three states, there is a confusing aspiration of certain Afar armed groups, struggling for a unified Afar homeland. This aspiration is often looked at with great care and suspicion by the three states. In effect, they collaborate to liquidate any group harbouring the aspiration of a unified and independent Afar homeland. This was demonstrated in the tripartite campaign of the three states to crush the Afar rebellion against the Issa-Somali dominated government of Djibouti, led by the Front for the Restoration of Unity and Democracy (FRUD) in 1992. The FRUD mobilised all Afar rebellion and galvanised popular support, threatening to overthrow the Djibouti government. The armies of Eritrea and Ethiopia came to the rescue of the Djibouti government in time to halt the FRUD insurgents from advancing on the capital of Djibouti (Schrader, 1993; Yasin, 2008).

According to an informant who was one of the leaders of the Ethiopian Afar battalion that joined the FRUD rebellion, the concerted effort of the Afar Horn states against the FRUD rebellion in Djibouti was a sign of their common antipathy to any form of secessionist movement. He argues that the idea of a united Afar homeland is considered a great threat to all Afar Horn states against which they will mobilise and join forces (Idris, 2011). Eritrea opposes all Afar movement because of the threat of Red Sea Afar separatism, which is supported by the Ethiopian and Djibouti Afar. During the independence of Eritrea, the separation of the Southern Red Sea Afar from Ethiopia was vehemently protested against by mainly the Ethiopian Afar led by Sultan Alimirah Hanfere. Hence, redrawing the border was meant to sever the strong solidarity of the Afar people, and strategically redefine their relations. This was a mechanism of monopolising the transnational allegiance of the Eritrean Afar under Eritrean citizenship. The EEBW was meant to do the business of breaking the transnational allegiance of the Afar on behalf of Eritrea, and manipulate the grievance of Red Sea Afar, to rally them against the Eritrean government on behalf of Ethiopia. During the border war and thereafter, both states were engaged in the classical strategy of politicking in the Horn of Africa: promote mutual insurgency to destabilise one another. Afar insurgency, situated in the contentious borders and highly securitised borderlands of the Afar Triangle, was caught in the crossfire as the cannon fodder of inter-state struggles (Muauz, 2016).

According to a UNSC Somalia–Eritrea monitory group report, Eritrea has continued its agenda of destabilising both Djibouti and Ethiopia:

> Eritrea continued to provide support to armed groups intent on destabilizing Ethiopia and Djibouti, including the Benishangul People's Liberation Movement, the Front for the Restoration of Unity and Democracy (FRUD-Armé), Patriotic Ginbot Sebat (PG7) and the Tigray People's Democratic Movement. While none of these groups poses a critical threat to either Djibouti or Ethiopia, the support of Eritrea for them continues

to generate insecurity in the region and undermines the normalization of relations between regional Member States.

(UNSC, 2016)

Ethiopia, as part of its regional policy, upholds a mixture of irreversibility and reversibility of borders shared with Eritrea. After the bloody EEBW, Ethiopia and Eritrea agreed to an arbitration commission. Consequently, the Eritrea and Ethiopia Border Commission (EEBC) was established in 2001. The EEBC, which was made up of eminent legal experts selected by the two parties, was given the dual mandate "to delimit and demarcate the colonial treaty border based on pertinent colonial treaties (1900, 1902 and 1908) and applicable international law". Both parties agreed that the decision of the EEBC would be final and binding and would be followed by the expeditious demarcation of the border. The EEBC based its decision on the provisions of the above colonial treaties, and the effective administrative claim overriding the provisions of the colonial treaties (AU, 2000). The Commission also referred to the AHG/Res, 16(1) OAU Cairo Declaration of keeping colonial borders intact, which recognised the border of Eritrea upon independence (EEBC, 2002; UN, 2002: 85).

Based on the 1902 treaty, which provided that the Cunama could belong to Eritrea, in 2002, the Commission awarded Eritrea the western sector of the border, where *Badme* town is located. Accordingly, the whole Cunama territory was also awarded to Eritrea, invalidating Ethiopia's claim (EEBC, 2002). However, the decision left the exact location of the most sensitive town of *Badme* undetermined but indicated the coordinates of the border. This created confusion, making both parties claim victory. Unfortunately, because of the confusion involved in the decision and the involvement of sensitive and symbolic territories such as *Badme*, the decision of the EEBC was not implemented. Neither of the parties are ready to forgo territories like *Badme* based on a give-and-take negotiated settlement. The political cost of losing the principality of *Badme* in exchange for a lasting settlement of the border conflict for both the governments of Ethiopia and Eritrea, has remained too high to afford. The settlement of the less contentious and less sensitive Ethio-Eritrean border running through the Afar homeland, as the rest of the Ethio-Eritrean border, has been left undetermined, and continues to be a flash point of military confrontation. Consequently, caught in a no-peace-no-war state, the border and the borderlands of Afar on both sides of the divide have become the most volatile, militarised and securitised border in the Horn of Africa (Medhane, 2008).

Ten years after the EEBW broke out, in June 2008, the Ras Doumeirah incident happened. Eritrean forces inroaded the Ras Doumeirah principality, a strategic place at the narrowest crossing point to the Gulf of Aden at the strait of Babeal Mendeb. The locality is also known for being the home of Afar insurgency in Djibouti. After a couple of days' skirmish, the Eritrean armed forces easily controlled Ras Doumeirah, routing the Djiboutian armed forces. The incident pulled Ethiopian forces into the conflict in order to secure the crucial export and import life line connecting Ethiopia to the port of Djibouti after the closing of the Eritrean port of Assab. Djibouti took the case to the AU, the Inter-Governmental Authority on Development (IGAD), the Arab League and subsequently to the United Nations Security Council (UNSC). Against Eritrea's rejection of any conflict, a presidential statement (S/PRST/2008/20) on 12 June was issued by UNSC, condemning Eritrea's military aggression against Djibouti in Ras Doumeira and Doumeira Island.

The statement requested the withdrawal of Eritrean forces to the *status quo ante* (UNSC, 2008). The AU and the Arab League urged both parties to peaceful resolution of the conflict. Djibouti complied with the presidential statement and withdrew its forces to the *status quo ante*. Eritrea remained defiant to all peaceful efforts and to the resolution. Eritrea also rejected fact finding missions dispatched by the AU and the Arab League. As a result, the UNSC issued Resolution 1862 (2009) for the cessation of hostilities, and the withdrawal of all armed forces to the *status quo ante*. It also required Eritrea to refrain from destabilising the Somali peace process by providing support to radical Islamist forces like *Al Shabab* (UNSC, 2009). Eritrea denied any involvement in border conflict with Djibouti and in the internal affairs of Somalia. The Djibouti representative appealed to the UNSC about the continued threats from Eritrea to the national security of Djibouti, contrary to Djibouti's peaceful resolution of the dispute (UNSC, April 2009).

The UNSC issued resolution 1907(2009), imposed targeted sanctions on Eritrea, and put in place an arms embargo, that restricted financial and physical movements of Eritrea's top leadership. The sanctions' purpose was to end Eritrea's role of destabilising the region and to respect the previous resolution for peace (UNSC, December 2009). Eritrea maintained its defiance against successive resolutions and the sanctions. In 2010, Qatari-led mediation resulted in the signing of an agreement by both parties for peaceful resolution. Eritrean forces withdrew to the *status quo ante*. The disputed area was controlled by Qatari observers stationed along the border who were mandated to look for lasting solutions binding to both parties. The UNSC report on Eritrea appreciated Eritrea's readiness to comply with its international obligations and with the UNSC resolutions, expressed by the letter of President Isayas Afeworki (UNSC, 2010). Based on an IGAD report on the continued destabilising role of Eritrea and its failed attempt to launch an attack to disturb the AU summit in Addis Ababa, the UNSC called upon "Eritrea to engage constructively with Djibouti to resolve the border dispute and reaffirm[ed] its intention to take further targeted measures against those who obstruct implementation of resolution 1862 (2009)".

Despite Eritrea's avowal to respect UNSC resolutions, its destabilising role in involving itself in Somalia, and supporting insurgents against Djibouti and Ethiopia, has continued (UNSC, December 2011). After eight years of the Qatari observer mission in Ras Doumeirah and Ras Doumeirah Island, nothing meaningfully changed for the better. Qatari military observers and peacekeepers remained in the disputed territory from 2010 to 2017. In 2015, a breakthrough for a lasting solution was attempted, but failed because of Eritrea's lack of readiness. On 14 June, it pulled its forces out of Ras Doumeirah and Doumeirah Island, the same day as Eritrean forces made inroads into the territory. According to the news agency Reuters, "Qatar said it was pulling its contingent out on June 14, days after the two East African countries sided with Saudi Arabia and its allies in their standoff with Qatar" (Reuters, 2017a; BBC, 2017). Despite an AU message requesting calm and UNSC resolutions to be respected, Eritrea continued to occupy the territory (Reuters, 2017a). Consequently, Ras Doumeirah and Doumeirah Island remain flash points of potential armed conflict capable of reigniting Ethio-Eritrean border war. A UNSC monitory report on Somalia-Eritrea has compiled ample evidence and labelled Eritrea a threat to international peace and security (UNSC, November 2017).

Djibouti and Ethiopia uphold amicable and cooperative management of their boundaries and borderlands. This is evident in the Ethio-Djibouti Border Joint Committee (EDBJC). Djibouti's policy follows international arbitration based on the irreversibility of borders with Eritrea discussed above, and cooperative management of the border with Ethiopia. The EDBJC is part of a strategic security, military and economic alliance between Ethiopia and Djibouti that came into more prominence after the formation of the EEBC. Ethiopia became fully dependent on the port of Djibouti during the inter-war period with Eritrea, and it remains the major outlet for Ethiopia to the present. The EDBJC is composed of representatives of the federal and local security organs, detachments of the national defence force, customs police, national immigration, and administrative and security organs of the Afar National Regional State of Ethiopia and Djibouti. The organ is entrusted to ensure the safety and security of the Ethio-Djibouti tarmac road, the free flow of the busy traffic, exchange of military-security information, and the joint management of border control and migration. The committee meets quarterly to plan, evaluate and execute joint activities (Muauz, 2015).

This joint effort is also influenced by the recognition of mutual security concerns emanating from Eritrea and its destabilising policy. Therefore, the effectiveness of the work of the committee is curtailed, overladen with military security concerns of regional implication, restricting the effective management of borders and borderlands. Yet, as compared with other borders and borderlands, this section of the border in the Afar Horn is the most stable and manageable. However, located at the volatile triangular juncture of Afar Horn states, the use and utility of borders and borderland has been limited to state security concerns that overshadow its multidimensional utilities to the states (AAJSAB, 2010).

From the perspective of the borderlanders, the Afar Horn borders and borderlands have unique features. Borders are traversed by multiple structures of relations and communications enabling local borderlanders to cross highly securitised and militarised borders. Borderlanders often carry three identification cards produced by the three states, which they produce for border patrols at times when they are stopped. Mobility and exchange of goods, information and people are mediated by the presence of powerful social infrastructures of identity. These structures are rooted in age old trade routes, economic networks and clan structures, connecting the coast to the hinterlands of Ethiopia and Eritrea. The Afar borders are thus as much irrelevant as rigid structures as they are vital tools for life making. Marginalised, too far from the centres of power, and often considered as less governable areas of the state, the borderlanders turn these factors to their economic and social benefit. At times, states become clients of such local borderlanders who are hired as informers and relay stations of military-security information and missions (Saleh, 2012).

The expedient shifting of alliances, the manipulations of state security concerns, and the political economy of insurgency in all Afar Horn borderlands, characterise the region. Equally, there is an abundance of structures of violence and insecurity that in turn reinforces the securitisation and militarisation of borders and borderlands. The complex overlaps of state policies, fragile and porous borders, the concomitant securitisation of borderlands together with the resourcing of borders by borderlanders, engendered the crystallising of a misconception about borders and borderlands as zones of insecurity. However, a closer scrutiny of the capabilities provided by borders and borderlands as resources for the people at the periphery of states, reveals that there is an untapped potential for peace. This aspect will be discussed in the following section.

From the above discussion, the features of borders and borderlands in the Afar Horn can be summarised into the following major points: first, borders and borderlands are highly contested. Second, they are highly securitised and militarised, because of belligerent state policies, protracted Afar insurgency, migration, human trafficking, contraband, and a proliferation of small arms and light weapons (SALWS). All these add to the insecurity emanating from the geopolitical volatility of the Horn of Africa. Third, Afar Horn states pursue contradictory policies and strategies of the management of borders. Fourth, borders and borderlands are marked by a multitude of economic, social and political infrastructures, traversing state boundaries, and connecting peoples under different state systems and political constellations. These create the potential for cooperative engagement and peace, overlooked by states, but resourced by borderlanders and other non-state actors. Fifth, states, borderlanders and other non-state actors manipulate the contested nature of state borders. Finally, borders and borderlands are both irrelevant and relevant to the Afar people. Borders and borderlands are considered irrelevant because of the continuity of the united Afar aspiration which rejects the inhibiting role of state borders. This is relevant because the recognition of state borders provides ample opportunity for the resourcing of state borders and borderlands, and the interdependence of the Afar people.

The place of local communities and borderlanders in this general picture, however, remains dark, and influenced by the negative general feature of the nature of borders in the region. The forms of existence of borderlanders who live with the reality of the borders and who utilise them accordingly, has not been the focus of academic research and investigation. However, the borders and borderlands in the Afar Horn, while characterised by overlapping problems, have the potential for cooperative peace and development.

The trans-border trajectories of peace in the Afar Horn

The trans-border trajectories are treated here in three separate categories. The first belongs to the nature of state facilities located close to borders, economic infrastructures and the complementary interests of Afar Horn states. The second category is associated with the identity, way of life and customary institutions of the Afar people. The third category is imbedded in the ancient trade routes, local economic networks and historical connectedness of the Afar Horn states through the intermediary of the Afar homeland. The central argument of the discussion in this section is that, the Afar Horn states have failed to utilise the capabilities arising out of the three trajectories for peace and cooperative engagement for the improvement of peace and security in the sub-region.

The search for outlets to the sea has been Ethiopia's continued historical struggle. The secession of Eritrea and the resultant removal of access to the Red Sea coast, has made Ethiopia the largest land locked state on the continent, dependent on the port of Eritrea first and the port of Djibouti after the outbreak of the EEBW (Clapham, 2008). The remapping of the Afar Horn created a new relationship of interdependence among the three states. The ports do not have significant utility to any state other than Ethiopia. Also, the oil refinery of Assab in Eritrea had a similar utility to Ethiopia, cementing the symbiotic relations to all states. The port of Assab is located at the southern tip of the Red Sea in the traditional Afar Sultanate of Tio, and the port of Massawa is located in the northern zone of the Red Sea. Given the comparatively vast territory of Ethiopia, the port facilities of Eritrea and Djibouti can be considered part of the borderlands of Ethiopia, adjacent to the two states. Accordingly, the port of Assab had served as the major outlet to Ethiopia's trade until the outbreak of the EEBW in 1998. Both Eritrea and Ethiopia had benefitted from the services of the port and the refinery of Assab. After the war, the Ethiopian road to the sea shifted fully to the port of Djibouti. This has created a new power alliance between Ethiopia and Djibouti against Eritrea (Medhane, October 2008). Consequently, the port of Assab, and the refinery, were left unused. Alongside this, the capability for cooperative engagement, economic and military symbiosis in these facilities, were abandoned.

Two decades after the outbreak of the EEBW, Ethiopia's dependence on the port of Djibouti has exponentially increased with the country's fast developing economy. Added to this, the fast-growing population currently beyond 100 million, and the need to support the growing economy, the necessity of using the closely located port in the Afar Triangle (Djibouti, Assab and Massawa) has become immense. This is evident from the large-scale port facility expansion in Djibouti, the construction of the modern Ethio-Djibouti railway, the diversification of multiple alternative ports located in far places like Port Sudan, the port of Mombasa, and the ports of Hargessa and Berbera. Had it not been for the no-peace-no-war situation, the Eritrean ports of Assab and Massawa would have been preferable outlets to Ethiopia. Thus, Ethiopia has to diversify the use of distant ports in the Horn region that are relatively located (EBC, 2018).

In addition to the economic interdependence created by the regional ports, other trajectories of cooperative relations connect Ethiopia with its neighbours. These include, the construction of roads, electric power grid, railway lines, telecommunications and potable water infrastructure. Such symbiotic relations make the opportunity cost of war and animosity unfordable to all states. These are the aspects of trans-border trajectories not utilised in Eritrea's relations with Ethiopia and Djibouti. Furthermore, the vibrant trade and commerce along the route to the ports of Eritrea are additional benefits on all sides of the borders which have dissolved into thin air because of failure to utilise the strategic facilities. The border towns of Elidear and Assab have become ghost towns, while the towns along the Ethiopia–Djibouti road are flourishing and becoming vibrant urban centres of trade and commerce (Abdu, 2012; Muauz, 2010a). Failing to utilise these capabilities for cooperation, the border and borderlands have become a bone of contention and zone of confrontation among the states. The opportunity cost of animosity in maintaining the status quo has become insignificant which in turn exacerbates belligerence and the breakdown of other relations. The above case is exhibited in Eritrea's exit from the sub-regional organisation Inter Governmental Authority on Development (IGAD) located in Djibouti city, and its exclusion from the US led anti-terrorism coalition also located in Djibouti where Ethiopia has become a leading regional actor (CJF, 2008; also see Muauz, 2015: 235).

The other vital capabilities for peace which are not utilised by the states are located within the wide structures and system of Afar identity, way of life, culture and institutions. The Afar conception of identity, a unified Afar-ness which traces its genesis to time immemorial. The Afar believe in the myth of origin that they are the first to settle the earth[1] after the legendary camel which descended from the heavens to settle in the Afar homeland. The camel is the most important animal in the pastoralist Afar's way of life. Their economic, social and cultural functions revolve around camels. Written history of the Afar traces to a common ancestor known as Adal Maccis whose descendants had settled for thousands of years in the south eastern part of the Horn of Africa, mainly around the ancient port of Zaila. They are one of the ancient Southern Cushitic peoples who settled the Horn of Africa along with the Somalis (Gamaluddin and Hashim, 2004). They settled the Afar Triangle which covers (in the north) the Boori peninsula to the foot of Tigray and the Amhara highlands (to the west). In the east, the triangle stretches from Djibouti city along the railway line south, through Erer to Awash station. In the west, the two areas are joined together to form a vertex at Namal Fan, 75km northeast of Addis Ababa (Yasin, 2008).

Before the advent of colonialism and the Western state system arriving in this part of Africa, the Afar had established customary self-governing political systems of sultanates. These include the Sultanate of Adal, Awusa, Ba'adu and Northern Afar (Afar homeland of Ethiopia), the Sultanate of Tio and Baylul (Eritrea) and the Sultanate of Tajoura (Djibouti) which has been loosely part of the Ethiopian Empire state over the centuries. During the advent of the Ottoman and Egyptian expansion and later Western colonialism, the Afar homeland and the coastal territories of the Ethiopian Empire state were the frontiers of struggle for the survival of Ethiopian independence and the autonomy of the Afar people. The anti-colonial struggle which is coloured by a history of gallantry and vigilance, has consolidated the ancient conception of a united Afar identity and the sanctity and indivisibility of the Afar homeland, romantically called in the Afar language *Qafar-Baxxo* (Afar homeland). The unity of the Afar continued even after the conquest of the Afar coastal territories of Assab and Tajura by Italy and France, a subtle and classical non-violent manipulation of local sultans – the Sultanate of Tio and Baylul, and the Sultanate of Tajoura – by Italians and French respectively. Despite their past, both sultanates have maintained over-lordship and a united conception of the Afar homeland. The Afar sultanates in Ethiopia had maintained close ties with the formerly mentioned sultanates by warning them and organising resistance against the malicious intentions of the European to divide and destroy the united Afar homeland. One finds evidence of this in the Afar oral tradition and Afar poetry, transmitted from earlier times to the present day, by word of mouth (Muauz and Saleh, 2017).

Accordingly, after the advent of the Western colonial system, they struggled against the division in all possible ways. Afar nationalism as a modern phenomenon is connected to this anti-colonial resistance and commitment to the unity of the Afar homeland which vehemently rejected colonial borders. One finds this symbolised in the popular adage of Sultan Ali Mirah of Awusa, uttered in opposition to the division of the Red Sea Afar from Ethiopia during the secession of Eritrea, referring to the place of Afar nationalism in Ethiopian state nationalism: "Let alone our people even our camels know the flag of Ethiopia." The Ethiopian flag is the icon of a united Afar homeland (Muauz, 2015). This sense of unity was rooted in unifying political, social and economic structures of the Afar people.

The customary political institutions of the Afar people, organised in various sultanates as much as they were self-governing, were inter-connected by allegiance to a hierarchy of customary rules of engagement governing their relationships. This is based on the historically rooted superior status of certain sultanates in relation to different aspects of political, judicial and religious affairs. Side by side with the egalitarian social organisation of the Afar people discussed below, sultanates enjoy the status of the first among equals to whom allegiance is paid by the rest. According to Afar informants, the Sultanate of Awusa enjoys primacy of making political decisions involving issues of the Afar homeland, territory and inter-Afar relations. The other sultanates in Eritrea and Djibouti pay homage to the Awusa sultanate in Ethiopia. Sultan Ali Mirah Hanfere of Awusa used to play the figure head of all Afar. His successor Sultan Hanfere Ali Miraha Hanfere inherited the same power. The coronation of the Sultan of Hanfere ten years ago was colourfully celebrated by representatives of all sultanates in Aysaeta, the historical capital of the Awusa sultanate. Regarding issues of interpreting the Mada'a and officiating at religious ceremonies, the Sultan of Tio enjoys primacy over the rest (Abidu, 2010). The Northern Afar of Ethiopia are reputed to be a repository of historical narrative, interpretation of historical issues, oral history and folklore of the Afar people. The sultanate is also known for being the repository of 400 years of heritage made from precious stones and gold. During the celebration of the 400th anniversary of the sultanate in 2007 at Dalol, representatives of all sultanates participated, to express their respect for the unique role of the sultanate in the history of the Afar people. The most interesting aspect of the celebration was that the three national flags of Ethiopia, Djibouti and Eritrea were displayed to show the unity of the Afar people despite being under different political systems. The elders from Eritrea were silently allowed to cross the border to Ethiopia despite the animosity between the two states (Mohammed, 2007). This was a recognition of the unbreakable ties among the Afar people by the governments of Ethiopia and Eritrea. These vital Afar institutions which play a significant role in, and influence Afar life, are quintessential structures which can be used as inter-state trajectories of peace, unfortunately not utilised to date.

Another complementary structure uniting the Afar people is the system of clan confederacy and territoriality. The Afar social organisation is a clan-based egalitarian system. Various clans come together to form a clan confederacy united by a shared use of resources, territorial ownership and control, as well as security alliances. The various Afar clans organised into clan confederacies located in the three states and belonging to various customary political institutions, bear allegiance to, and solidarity with, their clan members settling across the whole Afar homeland. For instance, a Djibouti clan will have clan segments and customary leaders in Ethiopia and Eritrea, or vice versa. Therefore, clan members of Abile Haysemale in Ba'adu which is the lower Awash valley of Ethiopia, consider travelling to their clan territory in Djibouti as perfectly normal. The same mobility rule applies to the other Afar clans that range across state borders (Mohammed-Awol, 2011).

A remarkable case corroborating the above was observed by the author in 2012 at Semera police station when a group of young men from the Abile Haysemale clan of Ba'adu were brought in, captured while crossing the Ethio-

Djibouti border, accused of illegal migration. The youth were using the Afar customary poetic performative art and vehemently rejected the accusation of illegal migration, invoking their view of the border, that they were going to their own clan territory to visit their relatives. After five days of hunger strike and displays of performative art rejecting the accusation and throwing a counter accusation to their captors, that they were ignorant of the age old Afar conception of borders and territoriality, they were set free. This was a time when the Ethiopian government was carrying out an uncompromising campaign against illegal migration and human trafficking. The Semera police station was accommodating hundreds of illegal migrants, caught while attempting to cross the Ethio-Djibouti border (Muauz and Saleh, 2017: 23). The exceptional treatment of the Afar youngsters by the police, while detaining other Ethiopians accused of the same violation, is owed to their recognition of the customary conceptions of territory among the Afar which even state incumbents are supposed to respect.

This conception of territory is not limited to those who have immediate clan kin across the border per se. Rather, it applies to any Afar crossing Afar land. It is established by tradition that any Afar coming to an Afar settlement, wherever that territory may be located, must be open handedly received as brethren. According to an ex-member of an Afar insurgent movement known as *Uquggumo* roaming the borderlands of the Afar homeland, members of the insurgency and locals freely travel in spite of the presence of state borders (Date-Mussa, 2012). The invisibility of state borders, vanishing in the face of multiple activities of the Afar people who cross them, is a structural factor uniting these people. The connectedness of the Afar is further consolidated because of the trans-national allegiance of the Afar people to common Afar customary laws (*Mada'as* and *Mablo*) and customs (*Qadda*) of conflict transformation and resolution. The Afar people are administered by five customary laws belonging to the various regions in the three states. The *Mada'as* are known by the place where the laws were made and originated. The allegiance of Afar clans in Ethiopia could be to the *Meda'a* and its Judges (known as *Mada'a-Abbas*) either in Eritrea or Djibouti and vice versa. This makes various clans members of a customary legal system which has trans-national applications, effective and legitimate in the daily life of the Afar across state borders in the Afar homeland (Muauz, 2012). The meaning and validity of state borders is further challenged by the practice of free mobility during events of conflict resolution and reconciliation ceremonies. The social obligation of attending events such as marriage, religious and cultural ceremonies, and mourning rituals provide further context for cross-border mobility. Similarly, during periods of war and rebellion against respective states (the FRUD rebellion in Djibouti, the Issa–Afar conflict), Afar clans cross the borders to lend a hand to their kin. Regional mobilisation of Afar is thus commonplace (Tadesse and Yonas, 2002).

The above discussed cross-border mobility is imbedded in the nomadic pastoralist mode of production which is based on seasonal migration of cattle, and temporary settlements across multiple ecosystems. Mostly, the migration occurs within state boundaries, however, it is not exclusively limited to intra-state migration. There are also trans-national seasonal migration patterns extending deep into adjacent state borderlands. These are of two types: one is an age old seasonal migration pattern Afar that pastoralists in all of the Afar Horn states follow. For instance, the migration pattern used by Southern Red Sea Afar of Afambo sees them crossing to the northern zone of Afar in Berahile and vice versa (Muauz, 2015). Likewise, the seasonal migration routes mutually used by the eastern zone Afar of Ethiopia, and the Afar of Assab. The same pattern can be identified between the Awusa Afar and the Afar pastoralists from Djibouti. Afar pastoralists south of Assab, and the Afar pastoralists of northern Djibouti share the Ras Doumeirah localities as seasonal migration routes (Tadesse and Yonas, 2002).

The second is during periods of drought and famine. The Afar border-landers cross state borders to exchange supplies and seek subsistence. In case of the provision of emergency relief aid and support for refugees, border-landers cross borders to avail themselves of opportunities on the other side with no thought of the legality of such a migration. This has been the case during the recurrent drought and famine affecting the Afar pastoralists in all Afar Horn states over the past six decades (Lawrence and Mohaddin, 2004). These relations have made the borderlanders resilient and give them an adaptation mechanism against harsh conditions of drought, famine and periods of displacement from their settlements. Even during periods of abundance, the practice of sharing subsistence and providing gifts to relatives across the border is a common custom among the Afar. The practice of sharing and gifts is also another adaptation mechanism against periods of drought and famine and this strengthens community cohesion and sense of security (Hassen, 2010). This sharing-economy-across-borders is reinforced by the presence of ancient trade routes traversing the Afar homeland, and connecting coastal trade with the hinterland.

The Afar Horn is knitted together by a network of ancient trade routes. These include the trade routes which connect the centre of the Afar Horn states with the coast. The ancient trade routes have served various political centres in Ethiopia by being routes to the sea. Traditionally, the routes traverse clan territories where thousands of people serve as caravan guides, security, and providers of food and support. These trade routes are important arteries connecting the hinterlands with the coast and supplement the Afar people's subsistence. Remotely located, inaccessible to state border patrol and control, the routes are used for both legal and illegal activities, and serve to unite the Afar in the Afar homeland (Mohammed-Raya, 2010). It is worth noting that these trading routes are also sites for the activities of women. In the Horn of Africa, in general, and the Afar Horn, in particular, 60 per cent of petty traders who carry out their activities along, across or on the branches of these routes are women who depend on cross-border trade for their survival (UNDP, November 2016). In the Afar Horn women are involved in the petty cross-border trade as a way of ensuring food security. As compared to Somali women, the Afar women are not active in this regard. Only Afar women in urban centres actively engage in big informal cross-border trade involving huge capital. Relating to the potential of the cross-border structures of peace, the informal trade is an area where women are dominant.

However, a tentative assertion can be made on the imperative to take note of the gendered nature of insecurity related with cross-border trade in the Afar Horn and the need to add the element of gender in reimagining borders and borderlands as having the potential for providing regional peace. Prior research in peace-making efforts in the Horn of Africa have shown that the exclusion of women's agency and lack of voice have contributed to the failure of peace-making efforts (Lederach, 1999).

Another vital structure used by the borderlanders is the time-tested and effective information communication system known as Afar *Xaggu*. This a grassroots communication exchange mechanism used to transmit messages across the land about events and matters which affect the daily life of pastoralists particularly, and also sensitive military and security information. In short, the Afar *Xaggu* system is an integral part of the greater community and an important network of communication. *Xaggu* could be used for drought, conflict and other security related early warning systems among the modern states of the Afar Horn as it is used by the borderlanders in resourcing state borders and borderlands for their own benefit. *Xaggu* is the most reliable mechanism of information exchange and communication, and mitigates the distance and remoteness of clans. The use of this system of communication makes distance immaterial. The Afar way of life and pastoral livelihood is precarious and risky. For all purposes and utilities of the above structures, Afar *Xaggu* serves as the nervous system of the nation (Ahmed, 2010). For the borderlanders it is a way to override the existing constrictive conditions and utilise state borders and borderlands to their own benefit.

The various structures, networks and practices functioning as overarching trans-national trajectories, are vital mechanisms of survival and livelihood among the Afar in the Horn. The effective use of these structures for thousands of years has defined the unique pastoralist way of life of the people and inter-clan relationships. This continues in spite of their separation across state borders. There is little doubt that the tapestry of functions represents vital contexts and capabilities for peace in the region, provided the host states recognise this and utilise these unique conditions to nurture a common zone of peace in the Afar homeland in the interests of mitigating troubled inter-state relations among the Afar Horn states. Unfortunately, the host states are failing to recognise and capitalise on these opportunities for cooperative engagement. As a consequence, inter-state relations have remained hostile with the Afar caught in the middle. Consequently, in response to inhibiting conditions, the inter-state hostilities, borderlanders and other non-state actors have used the structures, networks and practices to maximise personal benefit to the detriment of security in the region. Therefore, the next section is devoted to the examination of the various trajectories of violence which grew out of the above stated capabilities for peace.

The trans-border trajectories of violence in the Afar-Horn

The trans-border trajectories of violence refer to those structures which contribute to, or are structures of, visible and invisible forms of violence, fear and insecurity, in the sub-region. It is important to distinguish between the positive and negative ways that the same structures of peace can be used by borderlanders and non-state actors, before discussing their use as structures of violence. One important distinction is to identify the violent and non-violent ways of resourcing borders and borderlands. Dereje and Hoehne (2008) have identified multidimensional ways of resourcing borders and borderlands in the Horn of Africa. This includes economic, political, identity, and rights and status, resources. In the Afar and Somali speaking Horn, the Issa Somali, as compared to their Afar counterparts in Ethiopia and Djibouti, have unparalleled mastery in the resourcing of borders and borderlands to secure local influence (Dereje and Hoehne, 2008: 10–18). However, Dereje and Hoehne, other than explaining the various resourcing mechanisms, have not depicted the structures as capabilities for regional peace, nor as structures of violence. Therefore, the following section will analyse the various structures of violence created during the resourcing of borders and borderlands, because of the failure of the state to utilise trans-national capabilities for peace.

Ancient trade routes are structures where old and new economic benefits converge and become resources for borderlanders and non-state actors in the Afar Horn. One form of new economic activity being carried out along those routes is the arbitrage economy created by the presence of state borders, proximity to the commodity supply chains, divergent economic conditions and policies, as well as price variations among the three Afar Horn states. The Afar of Eritrea, because of the closure of the once bustling port of Assab, the failing economic condition, and repressive state control of both agricultural subsistence commodities and manufactured goods, are on the receiving end of the arbitrage economy. For instance, in the period 2006–2012, the supply of *Teff* flour (a stable food item endemic to Ethiopia and Eritrea) and other agricultural consumer goods from Ethiopia were five times more expensive in the Assab Market than in Ethiopia. According to the Afar National Regional State Police Contraband Prevention quarterly report for the same period, the average value of subsistence commodities (contraband) confiscated monthly while on transit to Djibouti was 460,000 Birr. This amounts to money confiscated from contraband traders. It does not include the small subsistence tolerated by border controls and the huge amounts escaping their vigilance (Afar Police, 2011).

The supply of rice and other consumable commodities from Djibouti to Assab offered a 200 per cent profit margin. Similarly, the price of rice, sugar, cooking oil and spaghetti in Ethiopia, smuggled from Djibouti, is 100 per cent more expensive than in Djibouti (Anonymous, 2011). The Afar of Djibouti are actively engaged at both the receiving end of agricultural subsistence items like sorghum from borderland Ethiopian markets at a 50 per cent lower price than Djibouti, and supplying manufactured goods to Ethiopian borderland markets. According to the USAID Famine Early Warning Network (FEWS NET) assessment, proximate Ethiopian markets are the only response mechanism for borderlander Afar pastoralists (USAID-FEWS NET, 2009).

These complementary trans-border commercial interactions are vital coping mechanisms for the borderlanders existing at the economic and political margins of their states. It is hardly possible to judge these micro trans-border transactions as a source of insecurity. However, in the strictest sense of the law, these micro trans-border transactions are acts of contraband, defined as illegal by state laws. For instance, according to the Ethiopian anti-contraband proclamation, contraband is defined as "an illegal act of smuggling in and/or out, holding, storing, trafficking, distributing, re-importing of legally exported commodities or any attempt of smuggling in and out including cooperating with such acts in violation of rules and regulations determined by law" (FCA, February 2009: 2). The micro trans-border trade of pastoralists could be tolerable activities not requiring full and strict enforcement of the law.

The necessity of the proclamation is due to the grave security challenges associated with large scale contraband networks. This refers to the involvement of a well-organised and massive trans-border contraband network running along the same trade routes. According to the Ethiopian Customs Authority Assessment Report, millions of US dollars-worth of commodity is transported under a well-organised and armed escort equipped with state-of-the-art communication technology. The route follows mainly the Issa contraband corridor together with the Afar corridor, feeding the three well known contraband market towns of Adaytu, Gedamytu and Quandafuqo in the Afar region, all serving as the supplying centres to the north, centre and partly south western part of Ethiopia (FCA, February 2009: 9–12). The national security concern is exacerbated due to the involvement of human and arms trafficking along these contraband corridors and the extent to which the illicit networks are imbedded in the social, the identity and the geographic features of the Afar Horn. Illicit trade in general, contraband trade in particular, requires access to adjacent markets, passages to ports, and suitable territory and social structure, as well as geographic features, that makes state control difficult (Muauz, 2010b: 98). The enterprise of exploiting the identity, solidarity, customary protection and information communication systems of the Afar people for the contraband, human and arms trafficking, has turned the capabilities for regional peace into instruments of violence against the economic wellbeing, human fabric and national security of states.

This is aggravated as the various forms of violence, human and arms trafficking associated with contraband meld. One such network is reflected in the guy–violence–gun cycle and financing of communal conflicts in Ethiopia: the proliferation of arms enables the outbreak of violence, and in turn the perpetuation of violence requires a greater supply of arms. A case in point is the Issa–Afar violence in Ethiopia which is partly a matter of controlling contraband corridors and benefits. On the one hand, the lucrative contraband returns are used to finance the perpetuation of Issa–Afar violence to ensure the free flow of contraband trade; on the other, the supply of instruments of violence – SALWs – to the communities, inhibits the de-escalation of Issa–Afar violence (Indris-Wonbede, 2009). In short, what can be seen as a nexus has made the opportunity cost of continuing violence an affordable enterprise for those associated with the illicit networks.

Mention of the guy–violence–gun cycle and the image of masculinity it evokes, makes it imperative to point out something here. One important issue often evaded in the study of violence and security is the gender dimension of it. Recent studies on the new fringe pastoralism that underscores the emergence of a new form of pastoralism associated with structures of violence stretching from local to global networks in the Horn and Sahel regions showed that women are highly affected by the violence involved. Moreover, women are excluded from policy and legislative considerations devoted to the promotion of the life of fringe communities as much as they are not empowered locally at the grassroots level (UN-ECA, 2017). It is clear that war and violence affect women more than and differently from men. Their degree of vulnerability to the various sources of insecurity related to the structures of violence is so high that it can be said that it has a gender dimension to it. This requires further research to expose the gendered dimension of insecurity and trans-border structures of violence.

The role of the politics of Afar insurgency in securitising borders and borderlands by its own merit notwithstanding, its connection with contraband, and human and arms trafficking networks has contributed to the conversion of trans-border structures into structures of violence. First, let us briefly look at the nature of Afar insurgency and how it utilises cross-border structures for propagating armed resistance in the Afar Horn. Afar insurgency is characterised by an insidious, rhizomatic and cross-generational continuity of armed struggle, which works in varying ways. It continues in various forms, and levels of organisation and mobilisation, pursuing different goals. During favourable times the organisation and mobilisation take the form of all Afar regional movements such as the case of the FRUD in Djibouti. During periods of political crisis and factionalism, it continues as a shadow organisation led by popular figureheads leading small bands; the case of *Mohammoda-Uguggumo, Saleh-Quguggumo* and *Arab-Uguggumo*, successive heirs of the big movement, are cases in point (Yasin, 2008).

In the context of inter-state war and tension (e.g. during and after the EEBW), insurgent groups like the Afar Liberation Front (ALF) underwent factionalism, one faction to reconcile and stand with Ethiopian government, and the other faction vanishing into a shadowy existence, offering at times, modest support to the Eritrean government and at others, staying neutral. The *Uguggumo* underwent factionalism; *Mohammoda-Uguggumo* disarmed and joined Ethiopian formal politics in the form of the Afar Revolutionary Democratic Unity Front (ARDUF). The other factions followed each other to head into the embrace of the Eritrean government. Similarly, other Eritrean Afar insurgents like the Red Sea Afar Democratic Organisation (RSADO) were born during the war period to play the role of *Uguggumo* against the Eritrean government. Their political programmes vary from the establishment of a united Afar state of all the Afar people (the various factions of *Uguggumo* and ARDUF) to a vague articulation of "independence" and "autonomy" (Afar National Revolutionary Democratic Unity Front –ANRDUF – and RSADO) to greater democratic rights in their respective states (ALF and FRUD) (Muauz, 2016: 5–6).

They sustain their struggle in different modes. Some are agents of the old political modus operandi, mutual insurgency of the Horn states; others independent movements antagonising all states or assuming a shadowy existence resurfacing when conditions are favourable. They undergo organisational, ideological and strategic mutation, to survive across generations. The major factor for the sustenance of Afar insurgency is the geographic nature, culture, identity and demographic infrastructure of the Afar Triangle. Insurgents can freely travel, mobilising and recruiting followers in the Afar Triangle borderlands, protected and harboured by strong Afar solidarity and other Afar institutions as discussed. The high regard the Afar have for people of the wilderness (known as *Garbbo*, a synonym for insurgency) as people of valour, manhood and resistance to the centre, and the grievance against the centre, offers them, when needed, a recruitment bonanza (Muauz, 2016: 5–6). Their various modes of existence and continuity feed on using those regional capabilities for peace, turning the structures and their organisation into permanent structures and agents of violence. One finds evidence of this in the role of the Afar insurgent, in the mutual insurgency and destabilisation activities of Eritrea and Ethiopia, who are using the other's insurgent group in the aggression equation. Their contribution is part of the explanation for the ever-escalating poor Ethio-Eritrean relations.

Their active involvement in contraband, trafficking in SALWs and humans to finance their struggle, and as part of military strategy of provoking and frustrating the centre, adds to the reinforcement of systems of violence in the region. This is even worse, given their role and involvement in local conflicts, being used by local governments as bargaining chips for central governments. An example of this is reflected in the way the Southern Red Sea Afar zone structure deals with the government of Asmara, promising to liquidate the RSADO (Abrahim-ALF, 2012). Similarly, the Afar region in Ethiopia uses the endless peaceful micro-disarmament, demobilisation and reintegration programmes (DDR) of Afar insurgents – unfortunately followed by rearmament to insurgency – to win political favours of the federal government of Ethiopia (Muauz, 2016: 22). All these practices are manipulated to secure political and economic gains with the issues of border and borderlands, lucrative industries.

In a slightly different fashion, borderlanders play similar roles. According to an informant from the Ethiopian national defence force who worked along the borderlands, as regards the ambiguity of, and conflicting claims over, boundaries, borderlanders equally manipulate the presence of insurgent movements by balancing political loyalties to accommodate political and military infrastructure from both sides of the border (Anonymous, 2012). Hoehne's findings from the study on the Somaliland/Puntland borders, have shown the rise of businesses related to cross-border migration of people, the refugee industry and the rise of new businesses (Hoehne, 2006). The behaviour of state actors is similarly shaped to consider borders and borderlands, and the transnational structures (effectively resourced by borderlanders on both sides of the divide to secure their livelihood and adapt to harsh conditions) as zones of insecurity, which more often than not, are reacted to with heavy handed military action. Such treatment is not caused by the mere phantoms of imagination. There are imminent security threats involved in those manipulating the transnational structures of peace into violence. The central contention of this chapter is that associated insecurities notwithstanding, borders and borderlands are not inherent zones of insecurity and ungovernability to be securitised. Instead, the insecurities are the result of failure on the part of the Afar Horn states to use the same cross-border structures used by agents of violence for the improvement of the life of borderlanders and the establishment of zones of peace for inter-state relations. Therefore, the transformation of the trajectories of cross-border violence into trajectories of peace, requires an imperative to reimagine borderlands as spaces of shared benefit rather than as entry points of risk and vulnerability.

Borders and borderlands as capabilities for regional peace

Our understanding about borders and borderlands in Africa is dominated by the idea of the nation state and Western theoretical caveats. This has precluded an appreciative critical inquiry about the positive roles and capabilities African borders and borderlands can play in inter-state relations on the continent. The evidence from the Afar Horn is instructive to reimagine the topic. It is possible, without the need to descend into revisionist versus rigid conceptions of African borders, to maintain a discourse on the constructive utilities of borders. Moreover, borders can be reimagined without exaggerating both the negative and positive features African borders and borderlands.

Borders in the Afar Horn are areas of violent confrontation between states, areas of free reign for illicit networks and insurgents whose roles reinforce false realities. For example, the idea that borders are unmanageable, anarchic, insecure and inherently zones of violence and insecurity, ranging over pastoralist land, rendered insecure because of pastoralist "gun-culture" (Okumu, n.d.: 10). That the post-colonial African states cannot secure the full-scale applications of their laws without remapping the artificial and porous colonial boundaries or using a securitised approach to border management is a false dichotomy creating false reality. This false reality exists because, in the case of the Afar Horn, the big picture of the Afar homeland is not exclusively defined by scenarios of insecurities. Rather, the region and its people are knitted together by overlapping institutions, relationships, identities and infrastructures, all of which ensure a peaceful and cooperative togetherness, a solidarity and fraternity, in the face of state and non-state violence. The resilient capabilities sustaining the unity and solidarity of the Afar people are played down to amplify the negative realities of the region.

One must finally ask a critical question at this juncture: if the borders and the people are inherently violent and insecure, how do we explain the reality that the people contained within those borders maintain trans-border cooperative and harmonious relations, while the host states are locked in violence? The answer is simple: the assumption is false and is based on the experience of the states and those resisting their sovereignty. The states consider borders as zones of insecurity, because they imagine the borderlands as a space of negative sovereignty where state power is contested by various actors. In effect, facing a security dilemma, they amplify either the voice of separatism, or their own call for monopoly over the legitimate use of violence. The Afar Horn states – by attending to the two voices, which are not non-existent, but derivatives of an issue not considered by the states – consider the Afar Horn as a space of existential threat to the territorial integrity of their states, legitimising excessive military measures.

Thus, state actors securitise the region, amplifying insecurities at the cost of imagining peace capabilities. This in turn criminalises the authentic peace capabilities in the region, rendering their existence invisible. It, further, legitimises the use of the capabilities for violence by illicit non-state actors and simultaneously legitimises state violence. Add to this the belligerent nature of the states (in part the result of a lack of local imagination), and the images of violent trajectories dominate the big picture of the peace capabilities. Therefore, states and illicit non-state actors are trapped in a vicious cycle of legitimising violence. Similarly, the imperative for reimagining the borders and borderlands of the Afar Horn as a cross-border space of peace endowed with peace capacities and structures which can be viably used for both state and non-state purposes, is dominated by excessive securitisation and militarisation of the region.

Even while states recognise the existence of these capabilities, using them for military security purposes, they imagine them as effective infrastructures for violence which in effect precludes their effectiveness and success. The states cannot imagine the Afar reality and world as in a continuous space unhindered by borders, wherein transnational trajectories smoothly function, uniting people. Illicit actors, for the obvious reason of evading capture, are more conscious about state borders than state actors. Yet, they imagine the space as tied into authentic trajectories and use them accordingly. First, illicit actors defying state control, first imagine these capabilities the same way as the people do; second, they invoke the disregard of the states to the values and way of life of the Afar people, at times sharing the life of the people and sharing the benefits to gain their sympathy and trust; and third, they use the capabilities and structures for different purposes.

Therefore, the growth of violent structures out of the peace capabilities is ensured because of the power of imagining the borders and borderlands as conceived by the borderlanders, that is as the authentic way of understanding the nature and values of borders. While agents of violence recognise and use the local conception of borders, the states remain alienated to it. This is evident from the continuity of protracted Afar insurgency and the prevalence of other illicit actors in the region (who are equally agents of violence). The failure of the states to use these capabilities and structures to manage border security to the degree the insurgents and other illicit actors utilise them is proof that the state system of Afar Horn is lacking insight into the authentic governing rules, capabilities, structures and values of the Afar homeland.

At a continental level, sources of lack of preventive and administrative predispositions, orientations, resources, human and institutional capabilities, relationships and commitments of exclusive states only are identified as major causes of border insecurity in Africa (Okumu, n.d.: 10–11). This would situate borderlanders' communities on the violence side of the balance sheet. The approach to border security where borderlander communities are imagined as agents of violence to be patrolled and controlled, is based on an erroneous conception of the authentic identity of borderlanders and their reality. It is based on mistaking those who manipulate the peace capabilities of the region for violent ends for borderlanders. It does not identify or recognise the peace capabilities that enable borderlanders to sustain themselves against the odds, against those who imagine their world as violent structures (the states) and against those who share their conception of borders and their world but manipulate it for violent ends (contrabandists, criminal networks, insurgents and terrorists).

This is not to deny the fact that the socio-cultural, economic and identity structures as well as the people are passive factors and subjects of manipulation. Not at all, for the structures that determine their lives are synchronised with their day to day activities. Nor does it mean that the borderlanders are not partly to blame in allowing agents of violence to use their socio-cultural, economic and identity space for their own ends. They support them as much as they actively and passively deny support to state actors trying to impose their own rules incongruent to, and denigrating, their system and values. Their divergent response is in part a show of political resistance to their multidimensional marginalisation and repression by the centre; in part the result of pressing challenges to sustain the declining viability of pastoralist livelihood using alternative means of income. The lack of trust, cooperative relations, and disarticulation of values of the states and the pastoralist borderlanders has precluded a shared understanding of borders. Besides, the belligerent inter-state relations situate the people of Afar homeland in a tight corner of sticking to their own age-old values and way of survival. Agents of violence work to buy the favour of borderlanders, who are conceived as threats. Consequently, agents of violence enjoy protection under the customary structures of the Afar homeland. If this were not the case, no perpetrator of violence against their values could escape from their control. That is why both Eritrean and Ethiopian military intelligence and counter intelligence, and mutual insurgency operations are most frequent across the Afar homeland borders.

Conclusion: reimagining borders and borderlands as capabilities for regional peace

So, what does reimagining the Afar Horn borders and borderlands mean? Does it mean the vanishing of the borders of Afar Horn states, the dismantling of the states' border control institutions, passports, the dissolution of their citizenship and the fusion of the Afar people into a common enclave of all states ruled by Afar norms? No, it means the opposite: reimagining at state level is meant to realise greater safety and security, and cooperative and symbiotic relations among Afar Horn states based on mutual trust, harmonisation of policies and constructive engagement. At community level, it means the realisation of basic human needs, freedoms and values, including the peripheries in development policies, synchronising state border management and security policies, and making constructive use of their authentic structures and institutions, all of which have the power to command compliance across the Afar. In light of this, to make the Afar borders and borderlands a space of harmonised, integrated, complementary and cooperative relations among the Afar Horn states by utilising the shared structures and values of the people. In short, it means turning the Afar Triangle and the state borders from being a hub of confrontation into an inter-state, intra-state and trans-border zone of peace. Furthermore, it entails using this part of the Afar Horn as the entry point towards greater regional integration, which will provide for free movement of people, commodities and ideas, governed by shared rules, cognisant of local rules and enforced by local communities. This requires reimagining at a series of levels.

The Afar as citizen of different states united by values, norms, identity and trans-border structures should be reimagined to be part of overall state strategies and approaches to managing borders, as well as relations, with neighbouring states. This requires understanding their aspiration as a quest for the sustainability of their livelihood, descent and uncoerced way of life which enables them to reject the use of their peace capabilities for violent ends. It means considering the Afar as important demographic assets possessing infrastructure and networks to support cooperative inter-state relations in the interests of building regional peace, and greater regional integration. The Afar homeland should be reimagined as a space existing in a complex web of pastoral seasonal migration routes, trade routes and clan territories, containing communication relay stations and symbolic land marks to which they have special attachment. Stakeholders in the Afar homeland must stop seeing the region as uncontrolled space only suitable for illicit actors and criminal networks. Failure to do so maintains modus operandi that includes military solutions, national security and unsustainable border management strategies.

To reimagine the trans-border peace capabilities as vital infrastructure with which state institutions, operations and procedures can be built and function, requires reimagining beyond the immediate violent episodes and events making use of these structures. The root causes of contraband, human and arms trafficking, and insurgency across Afar Horn borders are located in major domestic (social, economic, political, military and legal) and foreign policy considerations rather than the favourable contexts of the Afar world. It must be recognised that addressing the root causes, besides inhibiting the manipulation of the peace capabilities and structures for violent ends, will increase visibility and promote the viability of peace capabilities for national and regional peacebuilding efforts. Afar Horn borders and borderlands must be a reimagined cross-border space at the centre of inter-state military confrontation, settled by close kin across three states who, effectively and peacefully, crisscross them for cooperative and symbiotic benefit. Strengthening the solidarity of these people, removing state coercion and manipulation, requires rethinking the borders and borderlands as future zones of effective enforcement of state laws and strategies that take into consideration, the unique values and rules they embody.

It is necessary to reimagine the success of illicit non-state actors as indicators of the failure of states to address the basic human needs, freedoms and rights of borderlanders who give illicit actors the favourable contexts for their operations. Even borders, borderlands and transnational peace structures can be readily manipulated in the prevalence of marginalised, oppressed and alienated groups such as sub economic Afar pastoralists surviving at the margins of society. The image conjured, of illicit networks, is hardly viable in the presence of state institutions and structures synchronised with and utilising socially and culturally accepted peace structures that connect people across the region. Therefore, it requires reimagining these groups not as independently self-sustaining agents of violence, but as opportunist actors who bloom in the environment of inter-state hostility. The presence of marginalised and aggrieved borderland dwellers, in the absence of democratic rule and responsive governance, provides suitable context to maintain the operations of agents of violence.

In sum, reimagining the Afar Horn at various levels will enable state strategies and policies to work with local realities as important building blocks for peace and not merely objects of border control and management. Commitment to use cross-border peace capabilities reverses the tide (structures of violence) and opens a channel for the establishment of zones of peace. This can become the launching pad for regional integration in the Afar Horn and beyond.

Note

1 The myth seems not merely coincidental with modern paleontological findings of the oldest most complete human fossils of *Australopithecus afarensis* (the most famous of which is known as Lucy) and Ardi, even older than the former, being found in the Afar homeland. This would tend to corroborate the Afar myth of origin.

References

AAJSAB. 2010. *Ethio-Djibouti Border Committee Report of 2010*. Semera, Ethiopia: Afar Administration Justice and Security Affairs Bureau Archive.

Abdu, M. 2012. On the Impact of the Closure of Ethio-Eritrean Border on Trade and Commerce of Borderland Urban Centers [Interview] (16 June 2012).

Abidu, H. 2010. On Afar Customary Political Institutions [Interview] (20 February 2010).

Abrahim-ALF. 2012. On the Nature of Mutual Insurgencey in the Afar Horn [Interview] (16 June 2012).

Afar Police. 2011. *Afar National Regional State Police Contraband Prevention Quarterly Report from 2006–2011*. Semera, Ethiopia: Afar National Regional State Administration Justice and Security Affairs Bureau Archives.

Agnew, J. 2008. Borders on the Mind: Reframing Border Thinking. *Ethics and Global Politics* 1(4): 175–191.

Ahmed, Y. 2010. On the Utility of the Afar Xaggu [Interview] (15 May 2010).

Anderson, J. and O'Dowd, L. 1999. Contested Borders: Globalization and Ethnonational Conflict in Ireland. *Regional Studies* 33(7): 681–696.

Anonymous. 2011. On Cross Border Trade [Interview] (20 August 2011).

Anonymous. 2012. On the Role of Borderlanders in the Mutual Insurgency in the Afar Horn [Interview] (12 July 2012).

AU. 2000. The Algiers Peace Agreement, signed by Ethiopia and Eritrea on 12 December 2000. s.l.: Au.

BBC. 2017. What Is behind the Tension between Eritrea and Djibouti? BBC News. www.bbc.com/news/world-africa-40340210 (accessed January 2018).

Berouk, M. 2008. *The Eritrea-Djibouti border dispute. Situation Report*. 15 September. Addis Ababa: Institute of Security Studies.

CJF. 2008. *Combined Joint Force Bulletin, 3*. Djibouti: CJF.

Clapham, C. 2008. The Road to the Sea: The Regional Politics of Ethiopia's Trade. In F. S. A. I. Taylor (Ed.), *Afro Regions in the Dynamics of Cross Border Microregionalism in Africa*. Uppsala, Sweden: Nordiska Afrika Institute, pp. 136–152.

Cliffe, L. 1999. The Regional Dimensions of Conflict in the Horn of Africa. *Third World Quarterly* 20(1): 89–111.

Date-Mussa, I. 2012. On Afar Insurgency and their Movements across Afar Horn Borders [Interview] (16 June 2012).

DerejeFeyisa and Hoehne, Markus V. 2008. *Resourcing State Borders and Borderlands in the Horn of Africa, Working Papers*. Halle, Germany: Max Planck Institute for Social Anthropology.

Durand, F. 2015. Theoretical Framework of the Cross-Border Space Production: The Case of the Eurometropolis Lille-Kortrijk-Tournai. *Journal of Borderlands Studies* 30(3): 309–328.

EBC. 2018. TN News Dispatch, 12 January 2018. Ethiopian Broadcasting Corporation, Addis Ababa.

EEBC. 2002. *EEBC Report*, paragraph 3.54. s.l.: s.n.

FCA. 2009. *Anti-contraband and Commercial Fraud National Strategy, Federal Customs Authority* (February). Addis Ababa: FCA Archive.

Galtung, J. 1969. Violence, Peace and Peace Research. *Journal of Peace Research* 6(3): 167–192.

Galtung, J. 1990. Cultural Violence. *Journal of Peace Research* 27(3): 291–305.

Galtung, J. 1996. *Peace by Peaceful Means: Peace and Conflict, Development and Civilization*. London: Sage Publications.

Galtung, J. 2000. *Conflict Transformation by Peaceful Means (the Transcend Method)*. New York: UN.

Gamaluddin [Ibrahim Khalil A-Shami] and Hashim [Gamaluddin Ibrahim A-Shami]. 2004. *The Afar Almanac*. Addis Ababa, Ethiopia: Tirat Publishers.

Hassen, I. 2010. On Cross-Border Social Solidarity of the Afar [Interview] (10 February 2010).

Hoehne, M. 2006. Politics and People Along and Across the Somaliland/Puntland Border: Perspectives of Individual Actors, Borderland Communities and Political Centers. Paper presented at the conference Divided They Stand: The Affordances of State Borders in the Horn of Africa. Halle/Saale, Max Planck Institute for Social Anthropology.

Idris, W. 2011. On the Nature of the FRUD Led Rebellion [Interview] (10 September 2011).

Ikome, F. 2012. *Africa's International Borders as Potential Sources of Conflict and Future Threats to Peace and Security*. ISS Paper no 233 (May). Pretoria: ISS.

Indris-Wonbede. 2009. On The Impact of SALWs Trafficking on Issa–Afar Conflict [Interview] (September 2009).

Lawrence, Mark and Mohaddin, Hadija. 2004. *Djibouti Livelihood Profile, 2004*. UNICEF- Djibouti and Republic of Djibouti, Joint Appeal – Response Plan for Drought, Food and Nutrition Crisis, 2008. Djibouti: UNICEF.

Lederach, J. P. 1999. *Building Peace: Sustainable Reconstruction in Divided Societies*. Washington, DC: United States Institute of Peace Press.

Lederach, J. P. 2003. *The Little Hand Book of Conflict Transformation*. Pennsylvania, PA: Intercourse, Good Books.

Lederach, J. P. 2005. *The Moral Imagination: The Art and Soul of Peacebuilding*. Oxford: Oxford University Press.

Lyons, T. 1996. The International Context of Internal War: Ethiopia/ Eritrea. In E. K. A. D. Rothchild (Ed.), *Africa in the International Order: Rethinking State Sovereignty and Regional Security*. Boulder, CO: Lynne Rienner Publisher.

MedhaneTadesse. 1999. *Ethio-Eritrean Conflict in Retrospect and Prospect*. Addis Ababa: Mega Printing.

MedhaneTadesse. 2008. *The Djibouti–Eritrea Conflict, Briefing Center for Dialogue on Humanitarian, Peace and Development Issues in the Horn of Africa* (October). Addis Ababa: Inter Africa Group.

Miall, H. 2004. *Conflict Transformation: A Multi-Dimensional Task*. pdf: Berghof Research Center for Constructive Conflict Management.

Mohammed, W. 2007. On the 4th Century Celebration of the Northern Afar Sultans [Interview] (25 June 2007).

Mohammed-Awol, I. 2011 . On Afar Clan Relations and Mobility Across Borders of the Afar Horn [Interview] (September 2011).

Mohammed-Raya. 2010. On Trade Routes and Commercial Activities Connecting Afar Horn [Interview] (21 January 2010).

MuauzGidey. 2010a. Field Note on Observation of Borderland Towns: The Case of Elidear Town. Personal collection.

MuauzGidey. 2010b. *The Afar–Issa Conflict in the post-1991 Period: Transformative Exploration.* Thesis for the partial fulfillment of MA in PSS, AAU, UPEACE-Africa Program. Addis Ababa, AAU thesis collection.

MuauzGidey. 2012. The Mada'a and the Mablo: Customary System of Conflict Transformation Among the Afar. Wollo University Annual Research Proceeding no. 2. Wollo University, Dessie, Ethiopia.

MuauzGidey. 2015. The Geopolitics and Human Security of the Afar Horn in the Post-Cold War Period. *African Journal of Political Science* 9(6): 225–253.

MuauzGidey. 2016. The Paradox of Micro-Disarmament, Demobilization and Reintegration of Afar Insurgents in Post-2002 Ethiopia. Conference paper presented at University of Pretoria Humanities Research Day held on 19 August 2016.

MuauzGidey and Saleh Mohamed. 2017. The Poetic and Politics of Afar Kassow. *Eastern African Literary and Cultural Studies* 3(1): 19–39.

Nugent, P. 2002. *Smugglers, Secessionists and Loyal Citizens on the Ghana–Togo Frontier.* Athens: Ohio University Press.

OAU. 1964. *Resolution of AHG/Res16 (1) Adopted by the OAU Summit in Cairo in 1964.* Cairo: OAU.

Okumu, W. n.d. Border Management and Security in Africa. A working paper. s.l.: n.p.

Reuters. 2017a. Djibouti and Eritrea in Territorial Dispute after Qatar Peace-keepers Leave. World News. www.reuters.com/article/us-djibouti-eritrea-border-idUSKBN1980EP (accessed 10 January 2018).

Reuters. 2017b. African Union Urges Restraint on the Djibouti–Eritrea Border Spat. World News. www.google.com/search?q=WWW.+Reuters+African+Union+Urges +restraint&ie=ut f-8&oe=utf-8&client=firefox-b-ab (accessed 10 January 2018).

Saleh, A. 2012. On the Role of Borderlanders in Inter-state Security Relations [Interview] (17 June 2012).

Schrader, J. 1993. Ethnic Politics in Djibouti: From the Eye of the Hurricane to "Boiling Cauldron". *African Affairs, Journal of the Royal African Society* 93(307): 203–221.

TadesseBerhe and YonasAdaye. 2002. *Afar: Impact of Local Conflict on Regional Instability.* Pretoria, South Africa: Institute for Security Studies.

UN, 2002. Reports of International Arbitral Awards. Decision Regarding Delimitation of the Border between Eritrea and Ethiopia. 13 April 2002, Volume 25 pp. 83–195. s.l.: UN.

UNDP. 2016. *Ending Need Indeed: Harmonizing Humanitarian, Development and Security Priorities in the Horn of Africa.* UNDP Africa Sub-Regional Strategic Assessment No. 2 Consultation Draft (November). Addis Ababa: UNDP.

UN-ECA. 2017. *New Fringe Pastoralism: Conflict, Security and Development in the Horn of Africa and the Sahel.* Addis Ababa: ECA.

UNSC. 2008. *Presidential Statement (S/PRST/2008/20)* (12 June). New York: UNSC.

UNSC. 2009. *UNSC Resolution 1862(2009).* New York: UNSC.

UNSC. 2009. *Letter Dated 6 April 2009 from the Permanent Representative of Djibouti to the United Nations Addressed to the President of the Security Council. S/2009/180.* New York: UNSC.

UNSC. 2009. *UNSC Resolution 1907(2009).* Adopted by the Security Council at its 6254th meeting, on 23 December 2009. New York: UNSC.

UNSC. 2010. *Report of the Secretary-General on Eritrea. S/2010/327.* New York: UNSC.

UNSC. 2011. *UNSC, Resolution 2023 (2011).* Adopted by the Security Council at its 6674th meeting, on 5 December 2011. New York: UNSC.

UNSC. 2016. *Report of the Monitoring Group on Somalia and Eritrea pursuant to Security Council Resolution 2317 (2016): Eritrea. S/2016/2317.* New York: UNSC.

UNSC. 2017. *Letter Dated 2 November 2017 from the Chair of the Security Council Committee pursuant to Resolutions 751 (1992) and 1907 (2009) Concerning Somalia and Eritrea Addressed to the President of the Security Council. S/2017/925.* New York: UNSC.

USAID-FEWS NET. 2009. *Djibouti Livelihood National Profile.* Djibouti: USAID.

Yasin, M. 2008. Political History of Afar in Ethiopia and Eritrea. *Africa Spectrum* 42(1): 39–65.

4 Border porosity and counterinsurgency in Nigeria

James Okolie-Osemene and Benjamin Adeniran Aluko

Introduction

Border security is a trans-national phenomenon particularly in this era of insurgency and global terrorism, which tend to undermine the efforts of security agencies. This comes against the background that, across the millennia, migration or seasonal movements of people have been a significant aspect of the human experience in space and time (Schiller and Salazar, 2012). The significance of well-secured and manned borders cannot be downplayed in the international political system. This is based on the fact that well-manned borders contribute to the making of a modern and functional state. They reflect order and portray the government as capable of ensuring territorial integrity. In fact, well-manned borders would be difficult to reconfigure and cannot experience instabilities the way porous borders do. This claim is informed by the fact that porous borders are routes of illicit activity that sometimes constitutes conventional security threats in states. As a result, border policies have proliferated amongst the comity of nations in recent times to the extent that states have been compelled to review their immigration policies in line with globalisation and current realities confronting humankind. The influx of criminal networks has prompted states to embark on the review of their immigration policies, which are aimed at addressing the problems arising from trans-border activities, some of which can be described as non-conventional security threats.

Although Ekoko (2004) asserts that in most cases, political boundaries are artificial creations by humankind, and that the territories of states have become landmarks that cannot just be redrawn overnight; notwithstanding, Cassese (2016: 2) posits that:

> borders are becoming more "porous" and "malleable": they are crossed, can disappear, grow stronger, advance, retreat and redefine themselves. Borders are, in a word, subject to strong changes, dictated by diverse needs, some in the spirit of closure and some in the spirit of openness (elasticity of borders).

Thus, the reconfiguration that occurs within territories through illicit networks weakens national borders. In particular, this occurs when the openness of borders becomes a source of threat to national security when such routes are utilised by non-state actors for their selfish interests. The transnational networks that proliferate are mobilised for different purposes. Indeed, some transnational networks usually have economic and criminal purposes to the extent that they engage in underground border trade, currency trade and other activities. Most borders in Africa are characterised by different illicit activities that occasion the rush for border crossing for the purpose of transacting business. Such can lead to cross-border risk, which is any undesirable consequence arising from such activities and interactions and can affect states and individuals directly or indirectly. Hence, counterinsurgency is an endeavour that state security providers adopt to enhance human and national security (Ashkenazi, 2013).

The emergence of Boko Haram insurgency in 2009 marked a watershed in Nigeria's political, social and economic history. It took the Federal Government of Nigeria (FGN) many years to understand the recruitment, membership and factors that sustain the operational efficiency of the Islamist sect. As a result of the insurgency, the upsurge in violence since 2009 made Borno State, Adamawa and Yobe States of Nigeria the hotbeds of Boko Haram insurgency which affects many rural people and those in towns like Maiduguri, Yola, Mubi and Damaturu. Although attacks started in 2011 with many deaths recorded (Mauro, nd), the insurgency by Boko Haram and its resistance strategy increased in scale several months after the declaration of state of emergency on 14 May 2013, precisely in December 2013, when members of the sect coordinated and launched a series of deadly attacks on police stations in Northern cities including Kanama in Yobe State. The attack on immigration offices in January 2012 shows that no security agency has been spared by the group. In September 2013, they had already killed over 50 students of the School of Agriculture, Gujba in Yobe State, while about 49 people were killed in August in an attack on a Mosque and several villages including Konduga and Malari in Borno State (Olojo, 2014). Bama, Buratai and many other towns in Borno State also recorded different casualties.

The significant number of fatalities in the intervention incidents by the government army manifested during their counterinsurgency operations that became more extreme after the imposition of emergency rule to the extent that approximately 185 civilians were killed in a military operation against Boko Haram insurgents in Borno in April 2013 (Afeno, 2014). These military operations were ongoing while the issue of porous borders was not adequately taken care of for improved border security. The media widely reported Boko Haram's video showing the Nigerian pilot it claimed to have beheaded in September 2014 (Gabriel, 2014); and a second video released to mock government's counterinsurgency operations in which security chiefs claimed the sect's leader Abubakar Shekau was killed. The video revealed that the leader stated "I am alive" (Sulaimon, 2014). Also, the Nigerian Senate on Thursday,

25 September 2014 approved the $1 billion external loan requested by President Goodluck Jonathan to procure arms for the containment of threat by insurgents in the North (Ugonna, 2014; *Premium Times*, 2014). Both the Nigerian Government and the Boko Haram signed a controversial ceasefire reported on 17 October 2014, with such terms as suspension of hostilities and release of hostages. However, a notable problem is that it is not always clear whether the insurgent group embraces any planned ceasefire or not due to the group's inability to use media to acknowledge accepting the truce. After the report, Abadam and Dzur villages were raided by the group (Gabriel, 2014). The ceasefire violation shows that the sect was not sincerely ready to end its daily violence in the communities and states on Nigeria's border with Cameroon. It all shows that the insurgent group cannot be trusted or taken seriously in any ceasefire. Earlier, on 28 September 2014, the counter-insurgency operation received a boost when the Nigerian Defence Head-quarters disclosed that troops scored a strategic victory against the insurgents during the battle of Konduga. Since the kidnapping of Chibok schoolgirls in April 2014, which placed the community on the global map of insurgency, the Boko Haram sect continued to launch attacks to the extent that the 2015 Nigerian presidential election was postponed by one month due to the volatile nature of the region. Before 2015, Boko Haram controlled over 21 Local Government Areas in Borno State, but the recovery of the Sambisa forest in 2016 made the group embark on a recruitment drive (Irabor, 2018). The February 2018 kidnapping of more than 100 schoolgirls by the insurgent group in Dapchi village, Yobe State, is an indication that state security providers need to rethink their strategy to sustain the success recorded in the recovery of Sambisa forest, which limited the sect's operational capacity and efficiency. Despite the establishment of various joint security operations to strategically de-radicalise the Boko Haram sect, insurgent violence still scars most parts of Nigeria's Northeast due to porous borders.

Against this backdrop, this chapter examines how the porosity of borders facilitated Boko Haram insurgency, with implications for counter-insurgency operations. Apparently, porous borders enable the complexities of counterinsurgency, considering how insurgents' safe havens are strategically located around border communities where winning the hearts and minds of the local population have become a daily routine to maintain momentum. The porousness of borders has become a threat to states, a situation that facilitates conventional and non-conventional security threats such as external aggression, insurgency, organised crime and trafficking in persons, terrorism, smuggling of arms and various commodities. With key informant interviews involving scholars and border experts, some Nigerian soldiers and immigration officers involved in counterinsurgency operations, and secondary sources, this qualitative study responds to the following questions. Who are the security managers/actors around the borders? Why has insurgency remained difficult to contain despite border security efforts?

The problem of insurgents is that most of their objectives or reasons they launch attacks against the state are not realistic and cannot be achieved through violence. This is further complicated by their strategy of attacking both security installations and civilians with impunity as well as the execution of perceived spies or enemies of their ideology. So far, the counterinsurgency makes the security situation in Nigeria's Northeast an international matter. If not checked through timely security rethinking and action, insurgents can continue launching attacks and if successful, re-establish control of some parts of a country's territory and set up parallel systems of government, as was the case in 2014, when the Boko Haram controlled parts of Nigerian territory the size of Luxemburg. In view of this, this chapter argues that proper management of porous borders by relevant authorities and stake-holders will make it easier for Nigeria to overcome the threats posed by insurgents whose capacity to launch attacks on security forces has manifested in their access to sophisticated weapons, and until recently, training and grand strategy around the Sambisa forest and Lake Chad Basin area.

Some conceptual considerations

It is noteworthy that borders are demarcations that delineate state or quasi-state territorial boundaries (Benham, 2010; Amadi, Imoh-ita and Roger, 2015). Borders are marked areas or physical lines that divide two countries and at the same time can serve as ports of entry from one country to another, which usually require policing by immigration and customs officers who demand for documents like visas and passports from migrants (Nail, 2016). The demand for documents indicates that borders are sites that facilitate the socio-spatial differences between states (Houtum, 2005). It has been stated that, while borders are boundaries for some, they are gateways for others (Rumford, 2006 cited in Missbach, 2014). The gateway identity of borders in this regard is usually assumed by individuals who intend to take advantage of borders like asylum seekers, refugees, smugglers, traffickers, insurgents or ter-rorists whose activities may undermine the sacred nature of the borders. One of the challenges of borders is that not all of them have approved points of entry and that is why some of them are not manned by the security agencies. This explains why some borders are porous and can easily facilitate the migration of different kinds of people into a country.

The notable attributes of a border are that it divides two counties and reg-ulates the flow of people or things across or away from a frontier. It is non-static and can be readjusted as a result of natural disasters, inter-state conflict, negotiation of territory, juridical repartitions of legal domains, economic reforms that change labour restrictions or trade barriers and production zones (Eselebor, 2010; Nail, 2016; Schnyder, 2017). This non-static attribute mani-fested in the inter-state territorial dispute between Nigeria and Cameroon in which the International Court of Justice (ICJ), on 10 October 2002, ruled that Bakassi peninsular belongs to Cameroon and readjusted some boundaries between the two countries in the Northeastern part of Nigeria (ICJ/603/2002). The foregoing also points to the fact that the sustainability of a border depends on the efforts of government in regulating the activities of people through border management or maintenance.

The notion of counterinsurgency

The threat posed by insurgency endangers the ability of a country to develop and protect itself, and this can escalate if negotiation-focused peace agreement is not signed by the belligerent parties (Imobighe, 2001; Panwar, 2017; Sanchez and Illingworth, 2017). Insurgent troubled states are characterised by negative peace which manifests in state fragility and deployment of security forces. It is based on this need for self-actualisation and existence as a state that counterinsurgency is embarked upon. The US Government *Guide to the Analysis of Insurgency* (2012: 1), describes insurgency as "a protracted political-military struggle directed toward subverting or displacing the legitimacy of a constituted government or occupying power and completely or partially controlling the resources of a territory through the use of irregular military forces and illegal political organizations". The report further explains that counterinsurgency refers to "all the measures undertaken by a government to defeat an insurgency, including political, security, legal, economic, development, and psychological activities to create a holistic approach aimed at weakening the insurgents". The foregoing points to the fact that various agencies of the state must be carried along in the measure that is aimed at mobilising human and material resources by stakeholders in the military, paramilitary, ministries and civil society in order to contain an insurgency (Rineheart, 2010). According to Ibrahim, Kale and Muhammad (2016: 56):

> counterinsurgency is a campaign developed in balance along three pillars: security, political and economical, in accordance with Kilcullen's assertion "unity of effort" greatly depending on a shared diagnosis of the problem (i.e. the distributed knowledge of swarm) platforms for collaboration, information sharing and deconfliction.

The essence of unity of effort is to harness the required inter-agency human and material resources in fending off threats from the insurgents. Counter-insurgents can have a mandate to embark on operations under the aegis of a coalition involving different countries, either regional or allies with common interests. At this stage of counterinsurgency, winning the minds and hearts of the populace must be the priority of the government. One of the reasons many insurgent groups target the local population is to intimidate them or punish them for exposing their antics or informing government about their safe havens and modus operandi. Mobilisation, recruitment and remobilisation remain the strength of insurgents (Irabor, 2018). In every counterinsurgency operation, the ultimate goal is to annihilate the insurgent group members by every means possible, especially after the insurgents have declined negotiation efforts or windows of opportunity to surrender and embrace amnesty. The government cannot fold its arms and watch the insurgents continue with their campaign of creating a state of anarchy to undermine the rule of law.

Counterinsurgency also involves the recruitment of security operatives (in both the military and paramilitary forces), training, security education, weapons handling, intelligence sharing, communication oversight, grand strategy and proper understanding of military environment/strategic places including the strongholds of insurgents, deployment and redeployment of officers. Every counterinsurgency operation is expected to put a mechanism in place to discourage mutiny, which can downplay the military strategy. A counterinsurgency plan must incorporate offensive, defensive and stability operations in order to defeat the insurgents, establish government control and sustain peacebuilding and order (Albert, 2017). The desired results of these phases of counterinsurgency are difficult to achieve as long as the problem of porous borders is not contained. The fact that successes have been recorded in offensive and defensive operations means neither that the insurgents been totally defeated nor that communities are safe from reprisal attacks by the insurgents.

Idachaba (2017) has identified a source of concern in Africa where he sees strategic coordination as a crucial factor in counterinsurgency operations that is hindered by poor operational and counterinsurgency policies. The nature of a security system that is usually practised through intra-African cooperation in the form of joint security operation is crucial in the protection of territorial integrity of states (Imobighe, 2001). Therefore, in the spirit of a common security system, this security arrangement would equip them with a common front in countering the merchants of violence who are mostly outlaw security providers that have the propensity to promise the local population safety if they do not to open their hearts to the government agencies. As noted elsewhere (Okolie-Osemene, 2016; Okolie-Osemene and Okolie-Osemene, 2017), the activities of non-state armed groups must be checked through timely security sector reforms and the strengthening of the early warning system to make the job of counterinsurgents less complex. This will prevent the occurrence of ugly trends where communities or security forces are taken unawares by insurgents, like the situation in Philippines where Islamic militants and indigenous people's armed groups engage in beheadings and genital mutilation of Filipino soldiers despite government's counterinsurgency efforts (Fabe, 2016). The early warning will reduce the ambush of security forces and night raids in communities by insurgents who leave their targets and victims helpless. Troops have to apply intelligence in their bid to liberate communities and civilians held hostage in order to reduce casualties during operations.

Theoretical framework

This study is anchored on two theories, namely the political theory of territory and the theory of ungoverned spaces, which emphasise how poor security governance in a territory can facilitate the activities of non-state actors to the detriment of the state security providers, with implications for human and national security. Moore's political theory of territory which identifies groups as crucial stakeholders in territorial right-holding, explores the control over borders and resources within a geographical domain (Moore, 2015). When there is insecurity or contestation, people's moral right of occupancy which gives the basic right to live in a place, or moral entitlement to live there would be threatened (Moore, 2001). The risk of insurgency can displace people. It is noteworthy that ineffectiveness of states is liable to lead to spaces of a territory coming under the control of non-state actors, with implications for the state's territorial right; and any group able to impose order or demonstrate effectiveness could gain territorial right in a given area, which raises the possibility of a "might is right" argument that is contrary to international law (Moore, 2015; Taylor, 2016). Since the border is a territorial issue, this theory is apt because insurgency in Nigeria's context goes beyond criminality to involve the acquisition of territory. In essence, counterinsurgency is aimed at recovering the parts of a country's territory used by the insurgents as a base.

In this chapter, ungoverned territories refer to situations of failing, failed states and poorly controlled land, airspace or maritime borders which require state control (Taylor, 2016). Clunan and Trinkunas (2010) assert that the concept of the ungoverned space is inherent in doctrines on counterinsurgency, counterterrorism, counternarcotics, stabilisation, and reconstruction and peace building. Ungoverned spaces are physical areas which are social, political and economic zones where states do not have effective control, which non-state actors can exploit to avoid state surveillance and undermine state sovereignty (Clunan and Trinkunas, 2010; Raleigh and Dowd, 2013; Taylor, 2016). This weak capacity of states to make representation aggravates national identity crises (Idehen, 2016). Olaniyan (2017: 5) offers more insight into the complexities of ungoverned spaces:

> If spaces are effectively under the control of a functional state, then violent non-state actors (VNSAs) will never have a foundation for their occupation and there will be no basis for competition. Past and present examples of VNSAs successfully occupying a space and establishing some form of government show that these spaces were legitimately ungoverned, otherwise this need would not have been created and subsequently fulfilled. No VNSA can occupy legitimately governed spaces.

The foregoing exposes the weakness or failure of affected states in managing their territories. As a source of risk, the lack of governance over certain spaces also raises concerns over development, including the health, education, human rights and economic welfare of affected populations (Hoisington, 2013). This shows that ungoverned spaces aggravate regional and global insecurity with numerous non-state actors contesting state authority, because experience has shown that "non-state actors operating from ungoverned spaces pose challenges to the conventional paradigm of state-centric responses" (Clunan and Trinkunas, 2010: 276). The governance of ungoverned spaces is problematic because security lapses make non-state security providers merge as the alternative stakeholders of security provision.

The existence of an ungoverned space within a territory can simply be described as an ironic situation because a state is supposed to exercise full control within its national borders. Olaniyan (2017) argues that sporadic presence of security forces as a policy response to ungoverned spaces has not been helpful in Nigeria's efforts to address the problem of insecurity. For instance, the Sambisa forest, one of the several forests that constitute the ungoverned space in Nigerian geography, was not easily accessible to the Nigerian security agencies whose reluctance to storm the space has been identified as a product of their inefficiency, sub-standard weaponry and conspiracy (Rufai, 2017). The fact that ungoverned spaces are governed by elements that do not owe allegiance to central authorities of an existing state raises a serious security question (Olaniyan, 2017). This is because traditional or outlaw security providers can fill the gap to the detriment the state authorities. A significant percentage of cattle rustling operations are carried out in ungoverned spaces like unguarded forests, such as the Fagore, Kamuku and Kiyanbana in Northern Nigeria where Boko Haram insurgents capitalised on security lapses to unleash terror on people, including herdsmen (Rufai, 2017).

From the foregoing, it can be said that many porous borders can be described as ungoverned spaces through which insurgents easily execute their nefarious activities. Ungoverned spaces explain why most crime control model strategies in counterinsurgency suffer setback. The gap created by ungoverned spaces motivated safe havens which require counterinsurgency to guarantee security provision across borders. This shows that state authority is either non-existent or weak. This made the Nigerian government to deploy Operation Lafiya Dole in 2016 to establish security presence in ungoverned spaces.

Border challenges and issues in Nigeria's counterinsurgency operations against insurgents

Cross-border flows and insecurity

Porous borders show how cross-border migration flows affect state interests in various areas of national security, particularly that of state territorial integrity (Adamson, 2006). The border is a phenomenon that relates to the territories of modern states in the international political system, and the manipulation of borders by criminal networks is a pointer that even the authority of states to control parts of their territories can be threatened. One of the consequences of this development is the existence of "borderless" borders which are now easily accessed by insurgents. These are colonial borders inherited after the independence of many African states which are porous due to poor demarcation, thereby easily facilitating the illicit activities of criminal networks without interdiction (IFRA-Nigeria, 2013).

Given that security governance in such places is usually sporadic, preventing and detecting illegal entry through the border becomes a problem. Panwar (2017: 973) asserts that "non-state armed actors often operate across state borders without state control." Through actions that refuse the border aimed at creating a governed space scenario, contestation becomes more likely between state officials and non-state actors (Key Informant Interview, Melissa Schnyder, 2017). "Borderless" borders are spaces that experience flows or rapid movement of people with little or no stop and check activities; and this portrays a situation where the borderland which marks sides of the boundary between two countries is exposed to immigration crisis and illicit activities due to inadequate immigration. By implication, although there are clearly defined lines dividing the countries, the ungoverned nature of the location rather makes it seem "borderless" with unrestricted or unregulated movements. The eradication of "borderless" borders is therefore the responsibility of the countries that share the borderland.

Porous borders concern immigration in states. Adamson (2006: 166) has noted that "migration and migrants have a long history of being viewed as closely linked to national security concern". This caution cannot be disregarded considering how many states now make policies on how to regulate the movement of people and categorise legal and illegal migrants. By implication, such security concerns have made many states map out measures to strengthen their security sector's capability in countering any threat to territorial integrity by ensuring that the armed forces and the paramilitary are equipped with the needed skills and weapons to secure the territory. This is necessary because any state that does not address the problem of porous borders can have the challenge of state fragility. It demands a state must begin to prioritise migration policy formulation to reduce the possibility of having a contested territory.

The problem of ungoverned spaces has not been addressed by Nigerian policymakers. The persistence of ungoverned spaces continues to expose the weakness of state control (Idehen, 2016). Consequently, insurgents have had strategic advantage over security forces in places like forests around borders due to the inability of troops to hold the areas liberated without losing the communities to insurgents (Albert, 2017). The Boko Haram has benefited from the security lapses that led to the phenomenon of ungoverned spaces across the Lake Chad Basin countries which also contributed to the porous borders as seen in Nigeria's Northeast. The migration route attribute of the states across the Sahel explains why the authority of states is seemingly too distant from the people and the territories that policymakers are expected to protect. This environment of neglect is further worsened by arms proliferation which drives illicit trades that create the atmosphere of lawless enclave where other organised crimes like kidnapping and drug trafficking empower or embolden some of the actors to perpetrate the crimes (Weeraratne, 2017; Okolie-Osemene, 2017; Irabor, 2018; Okolie-Osemene, 2018).

Intra-state and inter-state collaboration

Nigeria has become a victim of what some scholars describe as state failure in the Sahel occasioned by lack of official policies for engaging with migrants (Norman, 2016). Consequently, weakness of states sharing borders in the area has been exacerbated by the various factors. These include instability, the collapse of Libyan security institutions, rebellion in Northern Mali by armed Tuareg groups, corruption in Sahel countries, activities of extremist groups, establishment of transnational criminal networks with smuggling routes for trafficking in weapons, drugs and people (Yamamoto, 2013). The porous borders have enhanced the activities of the insurgents more than before when they started their activities against the state. Arguably, Boko Haram insurgency is a reflection of the hollowness within Nigeria's overall security and aggravated by the failure of the state-building project in the country (Hentz and Solomon, 2017). This situation can be attributed to the multiple cleavages in terms of ethnocentrism, regional dynamics, cultural and religious differences. However, the escalation of the insurgency and attendant counter-insurgency has attracted multiple external actors.

The free movement of people has become a burden to communities that have unmanned borders, which brings to the fore the perception of "borderless" borders. It is evident that Nigeria's porous borders with countries like Chad, Cameroon, Niger and Benin Republic greatly aided insecurity, considering the porous nature of land borders encouraged by the Economic Community of West African States (ECOWAS) free movement of persons, goods and services (Ayegba, 2016). This is why inter-state cooperation is necessary because there are times when insurgents in crisis-ridden states intermingle with refugees in receiving countries with the aim of perpetrating nefarious acts, like a situation where fleeing rebels from Chad crossed into Nigeria (Ate, 2001; Albert, 2017). Consequently, until such persons are arrested, the town where they reside can become another hotbed of violence or insecurity. That is why many refugees must be screened for illegal arms while immigration and customs officials also need to embark on stop and search operations around border communities, forests and other isolated areas that may offer insurgents a safe haven to operate.

At the initial stages of the insurgency, the Nigerian government adopted a self-reliant strategy to curbing it but when it escalated, it became obvious that this was no longer a national security threat. Many efforts have been put in by different actors namely Joint Task Force (JTF), Nigeria Police Force (NPF), the Department of State Security (DSS), Nigerian Customs Service (NCS),

Nigeria Immigration Service (NIS), and the Defence Intelligence Agency (DIA). These agencies enforced the government's crime control model of national security. According to Animasawun (2013: 401), the JTF adopted stop and search operations, door to door security searches for weapons and sometimes killing of suspected insurgents as strategies for insurgent containment. Such approaches aimed at apprehending notorious members of the Boko Haram would have psychological effects on the people in the Northern region. Unfortunately, the government's initiative in fixing different price tags for relevant information from residents of Northern Nigeria to facilitate the arrest of Boko Haram leaders declared wanted could not de-escalate the violence.

The aim of countering these merchants of violence is to end humanitarian crisis, enhance food security and protect the vulnerable residents especially aged persons, women, children, disabled persons, IDPs (internationally displaced persons) and refugees. It is noteworthy that the activity of insurgents across the border has brought businesses to a halt in most parts of the Northeast that are near the borders. A Cameroonian who lives in a community near the border between Nigeria and Cameroon informed the author that life has not been the same since the escalation of insurgent violence and that starvation has been the order of the day due to inadequate food supplies from Nigerian communities ravaged by Boko Haram. It is evident that insurgency hinders social justice. Equating the movement of people across borders with crime has motivated mobility policies and programmes that are sometimes discriminatory to reduce movements of people, particularly in an era when porous borders and buoyancy of some economies relative to neighbouring countries attracts migrants from less developed countries (Oshita, 2010; Abram et al., 2017).

Prior to 2014, the weapons and bullet proof vests used by security forces did not help save the situation as some insurgents had weapons superior to those of the counterinsurgents. Security forces were shocked when residents mocked them (Interview with Immigration Officer in Borno State, 2016). The foregoing explains why the government had to source modern weapons across the world to enhance its security provision. For instance, in 2017, the Federal Government purchased Super Tucano A-29, an agile, propeller-driven plane with reconnaissance, surveillance and attack capabilities; and it is expected that the newly acquired 12 ground attack aircrafts (with sophisticated targeting gear) at the cost of $600m (*Premium Times*, 2017), will aid the counterinsurgency.

The extreme cross-border violence by Boko Haram also made member states of ECOWAS engage in intelligence sharing to frustrate the activities of insurgents, and their counterterrorism strategy that incorporates strengthening and harmonising their legal frameworks for a safer region (Clottey, 2015). During the fiftieth session of the ECOWAS Authority of Heads of State and Government in December 2016, they proposed the creation of a special solidarity fund aimed at assisting those affected by terror attacks and demanded the international community's support in the reconstruction of the Northeast communities ravaged by Boko Haram. It is noteworthy that, with the advent of armed conflicts and insurgencies, ECOWAS became a regional organisation that expanded its core objectives of economic integration to incorporate a security mandate. This was aimed at collaborating for a peaceful region in order to protect the territorial integrity of member states. By implication, non-state armed groups would find it difficult to establish safe havens in member states' territory because insecurity in any country in West Africa will eventually affect the others. This is the main reason why West African nations must establish joint security operations against insurgents with their counterparts in Central Africa. Such collaboration has good neighbourliness and multilateral relations value.

There are notable shortcomings in the security arrangement that involves the JTF. According to Nwozor (2013: 23):

> The setting up of Joint Task Forces is a de facto declaration of a state of emergency. Joint Task Forces have different rules of engagement, which are outside the normal operational boundaries of regular security agencies. Their mandate is a quasi-empowerment to engage in war. The code names often associated with specific Joint Task Force operations underscore its direction and strategy. Generally, the strategies of Joint Task Forces are anchored in their ad hoc composition and non-allegiance to any specific security agency. Their driving philosophy is shaped by the mindset that only superior force can tackle insecurity.

Today, the understanding between Nigeria and her neighbours that are members of the Lake Chad Basin Commission has yielded positive results as the presence of the Multinational Joint Task Force (MNJTF) makes it problematic for fugitive insurgents. Albert (2017) underscores the significance of Operation Gama Aiki by the Nigerian Air Force in finishing the job of routing the insurgents in the direction of the border with Niger and Chad where MNJTF troops are strategically positioned for action.

Porous borders made it difficult for the government to control the shipment of weapons and the movement of criminal elements (Shalangwa, 2013; Guéret, 2017; Albert, 2017; Weeraratne, 2017; Irabor, 2018) who infiltrated the security market. This automatically empowered the insurgents as merchants of violence who became the drivers of trans-border risks with their transnational networks, funding a growing number of sympathisers in various communities. The challenge of insurgency exacerbated by porous borders was also identified by then President of Nigeria, Goodluck Ebele Jonathan who raised alarm in 2012 that his cabinet and the security sector may have gradually been infiltrated by some sponsors and sympathisers of the Boko Haram insurgent group (Premium Times, 2012a). It is believed that this conspiracy frustrated most policies initiated to nip insecurity in the bud. Obviously, this was one of the reasons why the insurgency became difficult to contain despite various security operations and establishment of new barracks in the Northeast. Even taming the sources of Boko Haram's funding and weaponry became difficult despite their heavy deployment, aerial bombardment and discovery of some Boko Haram camps and bomb factories in the Northeastern enclave.

The factor responsible for the psychological strength and strategic advantage of the insurgents has been underscored by Lewis (2016) who asserts that insurgent group formation occurs in secrecy and in poorly monitored areas. And these are mostly areas such as mountains, valleys, forests, caves, uncompleted or abandoned buildings, among others, that are not easily accessible by state security providers. Forests are often used by criminal groups (including insurgents) as cover from which to launch attacks and as a defence against invasions (Olaniyan, 2017), such as that experienced when Camp Zero was invaded in December 2016 by the state security providers who are counterinsurgents.

The fact that most of the camps established by the two factions of the insurgents, Camp Zero and Camp Abuja, which aid netwar (a group's coordination of attacks without a precise central control command) against the government, are near Nigeria's border with Cameroon, with foothold advantage in the forests (Albert, 2017), should remind security agencies of the need to establish permanent presence in these locations even after the end of insurgency. This remains a necessity given that the destruction of Camp Zero in 2016 by the Rangers, senior battalion of the JTF (Key Informant Interview, 2018), which the Boko Haram leader Abubakar Shekau was using as his office, has not ended the insurgency unlike in civil wars where the taking over of a capital city or government house signifies victory or the end of hostilities.

The insurgency has contributed to the destruction of properties worth millions of dollars. Many bridges, homes, security posts, motor parks, markets and even schools have been damaged with bombs and other improvised explosive materials. A soldier even confirmed that when they recaptured most villages, the areas were already deserted, and they met nobody until two to three days later, when some villagers returned (Key Informant Interview, 2018). It is not disputable that insurgent violence has caused frustration by the security forces as a military jet mistakenly bombarded the Bama town's barracks and IDP camps on different occasions in a bid to rout the insurgents. This is caused by the conventional nature of the counterinsurgency compared to insurgents' adoption of guerrilla tactics, suicide bombings, kidnappings, hijacking and roadside shootings, among others.

The insurgents' activities, particularly ambush of soldiers and the destruction of various military installations, also contributed to the frustrations that motivated mutinies aimed at bringing to public knowledge, the suffering of troops. For instance, the counterinsurgency has the following unit risk factors: (a) high incidence of soldier and civilian deaths occurring in the same area of operation and over a short period of time; (b) the existence of enemies indistinguishable from innocent civilians (Sharma, Ratman and Madhu-sudhan, 2016). Mutinies recorded since the inception of counterinsurgency have occurred because sending soldiers to fight the insurgents with inadequate weapons is synonymous with sending them to die (Ibeh, 2014; Omor-otionmwan, 2017). This exposed poor military strategy in countering the insurgents, with implications for military values of discipline, loyalty and respect. Key informants involved in the counterinsurgency also confirmed their poor morale to the authors while conducting this study. According to Amao and Maiangwa (2017), the crossing of the border to Cameroon after an encounter with Boko Haram insurgents by over 400 Nigerian soldiers pointed to the problem of poor morale and poor-quality equipment for the troops. Moreover, the crossing of border communities made Cameroonian troops disarm the soldiers, although they maintained it was not intentional. Considering their sensitivity to spatial and territorial positioning, the tactical withdrawal that made them enter Cameroonian territory was done for safety in order to remobilise (Key Informant Interview, 2018; Okolie-Osemene, 2018). The poor equipping of the soldiers falls short of notable principles of counterinsurgency doctrine. Properly equipping them would have aided the operational needs of security forces involved in the operations.

It is on record that earlier on in 2012, Nigeria had already established bilateral cooperation with Cameroon in order to contain the Boko Haram monster through the signing of Trans-border Security Committee to engen-der the necessary political will in countering trans-border crimes (*Premium Times*, 2012b). The agreement yielded results as both countries initiated more meetings, particularly the bilateral meeting held in July 2017 between Nigerian military officers and their Cameroonian counterparts aimed at collaborating to curb the presence of insurgents around border communities (Olanrewaju, 2017).

For many decades, most parts of the Lake Chad Basin which now host insurgents have recorded population growth, budgetary problems of the states, structural adjustments, urbanisation, the crisis in pastoralist societies (notably the Fulani) and the influx of automatic weapons and battle-hardened men from vanquished armies in Niger's and Chad's wars (International Crisis Group, 2017). With automatic weapons at the disposal of insurgents, it is difficult preventing attacks on communities most of which do not have ade-quate protection of state security forces. The period between 2014 and 2016 witnessed the launch of multiple attacks on communities around the Lake Chad Basin due to the displacement of insurgents. The spatial nature of Boko Haram fighters has shown that insurgent violence is not restricted to only one geographical location and can easily spread from a town/country of origin to others. According to Pérouse de Montclos (2016: 2) a notable phase of recruitment was the "spatial expansion of attacks after the launch of an international coalition made up of Nigerian, Nigerien, Chadian and Camer-oonian armies in 2015". As seen from the foregoing, it suffices to assert that apart from insisting on engaging residents and examining the moral qualities

of interested youth, strengthening the membership criteria and initiating efficient means of screening will improve the reputation of vigilantes and prevent
infiltration of the groups by fugitive Boko Haram members who may want to
find out their strategies. Addressing the insider factor problem will enhance
community safety.

Vigilantism, insecurity and border communities

Vigilantism in Nigeria consists of support for extra-judicial ways of getting
rid of the inconvenient acts through the use of vigilantes which are social
organisations of private citizens set up to suppress deviance in the community
(Kowalewski, 1991; Sundar, 2010). For some time now, vigilantes have been
very effective in disrupting the ability of Boko Haram to execute attacks
(Stevenson, 2015). Their ability to fend off threats from the insurgents has
positioned them as security providers. Vigilantism in the Lake Chad Basin
(shared by Nigeria, Chad, Niger and Cameroon) is anchored on collaborative
strategies of stakeholders to address insecurity from the root. Vigilante groups
are beneficial to the people around Lake Chad because they are instrumental
in the identification and tracking down of the merchants of violence who are
not only the drivers of insecurity but also the enemies of the local population.
They also have the capacity to map out the safe havens of the criminals,
insurgents or bandits who utilise ungoverned spaces to unleash attacks on
communities. Vigilantes can identify hiding or fleeing insurgents and criminals who may want to mix with people in the community. Their knowledge
of the Lake Chad Basin terrain positions them to easily conduct a search
operation to expose the safe haven of criminals on the one hand and the
places where arms or kidnapped victims are hidden on the other hand. In
fact, engaging vigilantes in the task of monitoring movements makes it easier
for strange persons to be identified. As groups that work for the safety of
community members, vigilantes in the Lake Chad Basin area have made it
less frustrating for people to build community resilience, considering the
increased capacity to cope with risks of insecurity.

However, the challenges facing vigilantes are numerous. Gunmen use tri-
cycles and motorcycles to attack and abduct people. Also, some soldiers do
not trust the vigilante members despite their involvement in counterinsurgency (Key Informant Interview, 2018).

Vigilante members are now exposed to hazards due to the corrosive nature
of the Lake Chad Basin (Buchanan-Clarke and Knoope, 2017). The fact that
their relatives and families are already at the risk of internal displacement and
hunger makes it impossible for them to relax as criminals already have the
perception that vigilantism is a waste of time. Taking the battle to their enemies will help prevent looting. The informal nature of the vigilantes makes it
difficult for them to have the type of health provision and quality of life that
the security forces working for states within the Lake Chad Basin Commission do. This means that the survival of their families if they die while on

duty can be threatened despite the efforts they put into ensuring community resilience in this era of insecurity. Of course, the Adamawa and Borno State Governments have been helpful in motivating the vigilantes especially in areas of equipping them with needed weapons, information sharing and training to defend their communities.

Vigilantes have a pressing task of discovering locations that insurgents and other criminal gangs use as safe havens to launch attacks on communities and security forces. The hundreds of casualties recorded by vigilantes in different communities around Lake Chad (Kindzeka, 2015; Musa, 2016; Okolie-Osemene, 2017) show the risky nature of their task. In fact, most of them have been active in resisting attacks by thwarting the offensive tactics of the enemies of the people. The targets of vigilantes are not only insurgents, but armed bandits, cattle rustlers and looters who benefit from the displacement of people during insurgent violence (Cropley, 2017; Nigeria Stability and Reconciliation Programme, 2017). It then means that the targets vigilantes seem to respond to are multiple. The advent of insurgency and other criminal activities escalated the internal displacement and refugee problems which were not the attributes of daily life in the communities in the Lake Chad Basin, especially as a show of force between security forces and insurgents began.

The vigilantes have gradually advanced from using cutlasses, locally made guns/dane guns, daggers and other less sophisticated weapons to the use of modern firearms in response to the quality of the equipment of criminals and insurgents. The endorsement of vigilantes by some state governments has given vigilantism a moral boost as the idea is seen as a child of necessity as far as security around border communities is concerned (Key Informant Interview, 2017). The gradual recognition of vigilantes by some policymakers and the need to equip them with modern weapons to match the fire power of non-state armed groups has made them get support with modern firearms in order to complement the efforts of the security agencies already collaborating with neighbouring countries to block insurgents' routes across the borders used for heat and run attacks.

The insurgency caused by Boko Haram is a source of kidnapping of school girls and politicians in most parts of the region. The prevalence of extreme violence in communities made life brutish and motivated villages to adopt a self-help approach to human security. But the vigilantes have been facing the challenge of countering the insurgents whose operational strategies create fear in communities during the day and night. The vigilantes becoming security providers brought misery to communities in their bid to target the merchants of violence who carry out reprisals. This has been worsened by their use of weapons that do not match the sophisticated weapons used by the insurgents. Unfortunately, their effort is classified as informal legitimate violence which only complements the efforts of security forces. For the vigilantes, establishing footprints across all the communities affected by insurgency to reduce the scale of violence is a challenge as far as manning major roads to communities is concerned (Interview with Yake who now works with Search for Common Grounds, in Yola, 2017; Pérouse de Montclos, 2016).

An aspect of containing the insecurity occasioned by the insurgency in the Northeast is the remarkable initiative by the Nigerian military which embarked on an operation to win the hearts and minds of the people in Yobe, Adamawa and Borno through local militias known as *Kato da Gora* based on their knowledge of the terrain as well as their determination to eradicate safe havens of insurgents (Pérouse de Montclos, 2016). This is regarded as a civil-military collaboration to nip insecurity in the bud in the form of community based neighbourhood watch. It is also based on the fact that until the safe havens of insurgents are uncovered, every counterinsurgency or security restructuring will be futile. A notable challenge to winning hearts and minds is the incidents of human rights violations by the military as well as deaths caused by the bombings and gun violence targeting communities by insurgents who easily escape through some porous borders (Albert, 2017; Ahmad, 2018; Key Informant Interview, 2018). Although the Force Commander of the MNJTF, Major General Irabor has confirmed that the strength of Boko Haram is now limited to forests and islands, some border communities are still insecure, and this made the MNJTF consider policing the Lake Chad waterways to prevent the insurgents from regrouping in ungoverned spaces (Irabor, 2018).

To an extent, the strict adherence to human rights standards and ethical standards provided by the just war theory has prevented the military from destroying civilian populated areas suspected of harbouring some fugitive insurgents, with care usually taken by security forces to protect civilians (Albert, 2017). This does not mean that there is no killing of innocent citizens in the attempt to get rid of insurgents in communities. In fact, in January 2017, the Nigerian Air Force mistakenly bombed an IDPs' camp located in Rann, Borno State which is a border community towards Cameroon, after suspecting that insurgents were planning to regroup in Kala Balga Local Government. By the end of the bombing, hundreds of people including aid workers and IDPs had died (Joel and Akioye, 2017). The excuse that they received an intelligence report that insurgents were regrouping was not enough to embark on bombing without any attempt to either initiate surveillance or use ground forces to save lives. Even if the insurgents were disguised as IDPs, was there no better option than bombing the camp? This negates the mandate of a state security provider. This Operation Lafiya Dole's strategy of fighting the enemy from a distance had a lot of human costs with implications for Nigeria's human rights record. The images of the dead evoke memories of unprofessionalism and lack of intelligence, coordination or team work in Nigeria's security sector. It exposes the military's poor grand strategy as far as weapons handling is concerned.

The counterinsurgency in the Northeast has produced two groups of people, namely, those that are happy with the soldiers and the ones that have some of their children as Boko Haram members (Key Informant Interview,

2018). In terms of deployment length, over staying is affecting the mentality of most soldiers involved in the counterinsurgency to the extent that they are now very aggressive to each other (Key Informant Interview, 2018). This has been worsened by the fact that the long duration of deployment has separated soldiers from families (Archana and Kumar, 2016). Addressing this becomes critical to prevent misconduct behaviours by troops such as intoxication, suicide, desertion, mutiny and malingering, among others (Sharma, Ratman and Madhusudhan, 2016).

One may want to ask the question why it has become difficult to stop insurgents within and outside Nigeria where they operate from, despite various inter-state efforts to counter their activities. The reasons are not far-fetched considering how governments and other actors pay lip service to addressing the problem of unmanned borders where illicit activities take place. Nigeria's case shows that the actors involved in counterinsurgency did not have a grasp of how the insurgents operated until the insurgents had already taken advantage of the porous borders through which they brought in arms into the country.

The dispersal of the group is a complicating factor in counterinsurgency. The spread of insurgency to neighbouring countries is a manifestation of dispersal and fragmentation of the Boko Haram sect by the counter-insurgency, which has complicated attempts by the sect to maintain momentum in border communities around the Northeast region of Nigeria. Although the attack on their base at Sambisa was a threat as it made the future of the group bleak, destroying their artillery pieces remains a task that must be accomplished to guarantee victory. For the insurgents, it is apparent that the more they spread and gain influence in some communities, the more their sympathisers identify with them and complicate efforts aimed at routing them. Counterinsurgents face more challenges fighting insurgents when they are found in more than one community or state than when they operate in different towns across the borders of neighbouring states. The insurgency has made the communities around border areas experience human insecurity with attendant fatalities arising from the reprisal attacks on the residents who are now targeted by the insurgents in their bid to either establish their influence or attack security forces.

In terms of some security measures to tame the insurgency, the insurgency has necessitated the closure of Nigerian borders with some countries around the Lake Chad Basin. But it should be noted that closing already manned borders is no solution to the insurgency but priority must be given to various unmanned borders most of which are yet to be identified by the security forces. The International Refugee Rights Initiative observed in 2011 that African governments should not close their borders in order to refuse refugees admission, but emphasised the need for the African Union to examine the factors which cause civil conflicts, with a view to elaborate Comprehensive Plan Action for tackling the root causes of refugee flows and other displacements. It is evident that border closure or stringent security measures around national borders are aimed at preventing the influx of criminal groups and individuals that may easily become available tools in the hands of insurgents and recruited as fighters in volatile areas. In the case of the Northeast, the armed conflict has already created humanitarian crisis.

Porosity of borders undermines counterinsurgency operations in the Northeast as most insurgents easily launch attacks on troops and escape through the communities around borders most of which are unmanned, with implications for the control of arms. As part of the counterinsurgency strategy, arms control is critical to success. National Arms Control Strategy is apt, as scholars argue that "the strengthening of border security arrangement through enhanced intra and inter-state collaboration among security and intelligent agencies in the air, maritime and land border areas is vital in eradicating small arms" (Ani and Onyebukwa, 2016: 439). This is based on the fact that insurgency is aided by arms. The need to boost counterinsurgency and enhance border security led to presidential approval for the establishment of an operational base for the Nigerian Navy in Lake Chad Basin. This latest policy option by government aimed at taming the border insecurity demonstrates the strategic significance of the area to maintenance of Nigeria's territorial integrity considering how insurgents can exploit any security lapses if the area is not adequately policed. In fact, it should be noted that improving border security demands proper monitoring of maritime activities. The activities of insurgents who take advantage of porous borders have affected social cohesion around most parts of Nigeria's Northeast and neighbouring Niger, Chad and Cameroon with impact on sustainable livelihoods. Consequently, the Lake Chad Basin has remained a hotbed of extreme violence. This has aggravated the humanitarian crisis that affects vulnerable groups. The rising insecurity in this part of the country is a threat to human safety.

Conclusion

This chapter has examined the relationship between porous borders and insurgency which now affects Nigeria. Fuelled by the existence of ungoverned spaces, insurgency is a source of state fragility due to associated anarchy and extreme violence. With the political theory of territory and theory of ungoverned spaces which explain a relationship between groups and struggle for space in a geographical domain, the chapter has shown how the situation created unusual security concerns in the country with spillover effects in neighbouring countries that have their borders linking Nigeria. The actors of security and insecurity across the borders are the government and insurgents respectively. Border porosity has made counterinsurgency more difficult and challenging than expected considering the capacity of insurgents to coordinate attacks with sophisticated weapons, regroup and enter communities in border areas without much confrontation.

The withdrawal of insurgents and temporary stability in volatile areas should not be mistaken for victory against them. On the contrary, an area that has been liberated by counterinsurgency forces still needs a lot of security provision to avoid creating a vacuum. So far, some insurgents still launch attacks through Nigerian borders with Niger, Chad and Cameroon. Until insurgents are forced to surrender, are disarmed and are arrested or even killed, the insurgency is far from ending. This chapter maintains that porous borders made it possible for insurgents to infiltrate many communities to the extent that the insurgency has now become geopolitical and regional. The insurgency executed by Boko Haram can be contained by strengthening and sustaining Nigeria's surveillance efforts at the borders rather than by appointing only individuals that are regarded as efficient without a strong institutional approach to the problem. Based on the foregoing, there is no need for security chiefs to announce that victory is in sight within weeks or months.

The success of bilateral cooperation between Nigeria and Cameroon depends on these countries' ability to implement the components of the agreement they signed aimed at enhancing border security in defence of their territorial integrity. Taking the counterinsurgency beyond communities and cities to the borders is very critical to the success of the joint security operations. This is because the border areas are routes used by insurgents to transport weapons and fighters. Consequently, the idea of pushing the insurgents towards the border communities without properly clearing the border areas to guarantee territorial integrity of neighbouring states should be examined for a change of strategy particularly in setting up more mobile military and paramilitary checkpoints across the border areas.

The insurgency that was facilitated by porous borders has necessitated a community-based approach that must involve winning the people to the side of security operatives loyal to government. It is believed that through winning hearts and minds, military operations will gain the confidence of the local population as regards human security orientated military professionalism, and draw the people closer to the side of security forces. This would complicate the possibility of insurgents cooperating with idle youths to penetrate communities. The issue of winning hearts would be a mirage if the security forces do not demonstrate their readiness to work for the people, especially in more volatile areas. Achieving this will also require adequate protection of the local population from reprisal attacks by the insurgents for offering to work with the security agencies.

Enhanced border security will boost government's defensive and stability operations in communities that have been ravaged by insurgency since 2009. The earlier security operatives begin to identify the unmanned borders where insurgents gain entry into most villages in the Northeast, the safer the country's borderlines will be as far as routing insurgents is concerned. If achieved, the issue of insurgents establishing camps in border communities will become a thing of the past with appreciable impact on the prospects of counterinsurgency operations aimed at achieving the goal of protecting Nigeria's territorial integrity. It is therefore suggested that trans-border co-operations through a collaborative security management strategy should be prioritised by state security providers in their bid to make the borders safer zones. Also, apart from setting up insurgency emergency response systems around the border communities, it is timely for the government to encourage informal initiatives and engage with communities to counter incidences of insurgency and crime.

References

Abram, S., Bianco, B. F., Khosravi, S., Salazar, N. and de Genova, N. 2017. The Free Movement of People around the World would be Utopian: IUAES World Congress 2013: Evolving Humanity, Emerging Worlds, 5–10 August 2013. *Identities* 24(2): 123–155.

Adamson, F. B. 2006. Crossing Borders: International Migration and National Security. *International Security* 31(1): 165–199.

Afeno, S. O. 2014. Killings by the security forces in Nigeria: Mapping and trend analysis (2006–2014). Nigeria Watch Project, IFRA-Nigeria working papers series, no. 37. www.ifra-nigeria.org/IMG/pdf/killing-by-security-forces-nigeria.pdf

Aghedo, I. 2017. The Ubiquity of Violent Conflicts in Nigeria. *The Round Table* 106 (1): 97–99.

Ahmad, M. B. 2018. Restoring Peace in the Lake Chad Basin: A sine qua non for Successful Recharging of the Lake Chad. Paper presented at the International Conference on Saving the Lake Chad to Revitalize the Basin's Ecosystem for Sustainable Livelihood, Security and Development, organised by the Federal Government of Nigeria and the Lake Chad Basin Commission, 26–28 February, Transcorp Hilton Hotel, 1 Aguiyi Ironsi Street, Maitama, Abuja.

Albert, I. O. 2012. *Making Education a Force for Sustainable Peace and Development.* 25th–29th combined convocation lecture, 7 November, Ignatius Ajuru University of Education, Port Harcourt.

Albert, I. O. 2017. *Beyond Nigeria's Sambisa: Forests, Insurgency and Counterinsurgency in Africa.* Ibadan: Ibadan University Press.

Amadi, L., Imoh-ita, I. and Roger, A. 2015. Dynamics of Local Conflict in post 1990 Africa: A Case of Border Dispute. *International Affairs and Global Strategy* 36: 5–18.

Amao, O. B. and Maiangwa, B. 2017. Has the Giant Gone to Sleep? Re-assessing Nigeria's Response to the Liberian Civil War (1990–1997) and the Boko Haram Insurgency (2009–2015). *African Studies.* DOI: doi:10.1080/00020184.2017.1285665

Ani, K. J. and Onyebukwa, C. F. 2016. Nigerian Security Challenges and Recommendations for Sustainable Development. In P. I. Ukase, E. O. Akubor and A. I. Onoja (Eds), *Urbanization, Security and Development Issues in Nigeria 1914–2014, festschrift in Honour of Professor Enoch Oyedele.* Zaria: Ahmadu Bello University Press Limited, pp. 421–444.

Animasawun, A. J. G. 2013. Rethinking Hard-Power Responses to Radical Islamism in Bauchi, Nigeria. In I. O. Albert and W. A. Eselebor (Eds), *Managing Security in a Globalised World.* Ibadan: John Archers Publishers, pp. 384–394.

Archana and Kumar, U. 2016. Familial Pathways to Soldier Effectiveness. In Nidhi Maheshwari and Vineeth V. Kumar (Eds), *Military Psychology: Concepts, Trends and Interventions.* New Delhi: Sage Publications, pp. 283–297.

Ashkenazi, M. 2013. Forces of Order and Disorder: Security Providers and Conflict Management. In V. C. Franke and R. H. Dorff (Eds), *Conflict Management and Peacebuilding: Pillars of a New American Grand Strategy.* Carlisle, PA: United States Army War College Press, pp. 271–294.

Ate, B. E. 2001. The State System and African Security. In R. A. Akindele and B. E. Ate (Eds), *Beyond Conflict Resolution: Managing African Security in the 21st Century.* Ibadan: Vantage Publishers, pp. 57–73.

Ayegba, S. 2016. Interrogating the Performance of Security Agencies and Challenges of Prevailing Insecurity in Nigeria. In P. I. Ukase, E. O. Akubor and A. I. Onoja (Eds), *Urbanization, Security, and Development Issues in Nigeria 1914–2014, festschrift in Honour of Professor Enoch Oyedele.* Zaria: Ahmadu Bello University Press, pp. 478–522.

Benham, J. 2010. *Peacemaking in the Middle Ages.* Manchester: Manchester University Press.

Buchanan-Clarke, S. and Knoope, P. 2017. *The Boko Haram Insurgency: From Short Term Gains to Long Term Solutions.* The Institute for Justice and Reconciliation, Occasional Paper 23. 1–16.

Cassese, S. 2016. Territories, Peoples, Sovereignty. *Glocalism: Journal of Culture, Politics and Innovation* 3: 1–14.

Clottey, P. 2015. ECOWAS Implements Measures to Combat Terrorism. Voice of America, 2 December. www.voanews.com/a/ecowas-implements-measures-to-combat-terrorism/3084994.html (accessed 10 June 2017).

Clunan, A. and Trinkunas, H. (Eds) 2010. *Ungoverned Spaces: Alternatives to State Authority in an Era of Softened Sovereignty.* Stanford, CA: Stanford Security Studies.

Cropley, E. 2017. On Boko Haram Front Line, Nigerian Vigilantes Amass Victories and Power. Reuters, 15 June.www.reuters.com/article/us-nigeria-security-vigilantes/on-boko-haram-front-line-nigerian-vigilantes-amass-victories-and-power-idUSKBN1960FK (accessed February 2019).

Dowd, C. and Drury, A. 2017. Marginalisation, Insurgency and Civilian Insecurity: Boko Haram and the Lord's Resistance Army. *Peacebuilding* 5(2): 136–152.

Ekoko, A. E. 2004. *Boundaries and National Security.* Sixth Inaugural Lecture, Delta State University, Abraka, 25 March.

Eselebor, W. A. 2010. *Trends and Challenges in the Management of Security Borders in Nigeria.* PhD thesis submitted to the Institute of African Studies, University of Ibadan, Nigeria.

Fabe, A. P. H. 2016. Winning Hearts and Building Peace. In N. Maheshwari and V. V. Kumar (Eds), *Military Psychology: Concepts, Trends and Interventions.* New Delhi: Sage Publications, pp. 317–331.

Gabriel, C. 2014. Why FGN/Boko Haram Ceasefire Deal Failed. *Vanguard News,* 11 November.   www.vanguardngr.com/2014/11/fgboko-haram-ceasefire-deal-failed/ (accessed March 2019).

Guéret, T. 2017. Niger and the Fight against Violent Extremism in the Sahel. RUSI, London. 13 April. https://rusi.org/commentary/niger-and-fight-against-violent-extremism-sahel (accessed February 2019).

Hentz, J. and Solomon, H. (Eds) 2017. *Understanding Boko Haram: Terrorism and Insurgency in Africa.* London: Routledge.

Hoisington, M. 2013. International Law and Ungoverned Space. Available at: http://works.bepress.com/matthew_hoisington/4/ (accessed February 2019).

Houtum, H. V. 2005. The Geopolitics of Borders and Boundaries. *Geopolitics* 10(4): 672–679.

Ibeh, N. 2014. Nigerian Military Sentences 54 Soldiers to Death for Mutiny. *Premium Times*, 17 December. www.premiumtimesng.com/news/headlines/173470-breaking-nigerian-military-sentences-54-soldiers-death-mutiny.html (accessed February 2019).

Ibrahim, M., Kale, A. and Muhammad, A. I. 2016. Stemming the Tide of Insecurity Imposed by the Boko Haram Insurgency on the North East Nigeria: A Sociological View Point. *IOSR Journal of Humanities and Social Science* 21(12): 52–59.

Idachaba, E. 2017. Counterinsurgency Strategies and the Approach to Peace in Somalia: An Appraisal of the African Union. In S. Romaniuk, F. Grice, D. Irrera and S. Webb (Eds), *The Palgrave Handbook of Global Counterterrorism Policy*. London: Palgrave Macmillan, pp. 981–1002.

Idehen, R. O. 2016. Identity Crisis, State Vulnerability and Ungoverned Space in the Sahel Region of Africa: Implications for Nigerian National Security. *IOSR Journal of Humanities and Social Science* 21(9): 1–8.

IFRA-Nigeria 2013. Évaluation des risques au Niger et sur l'axe Niger-Nigéria. Rapport de synthèse. www.ifra-nigeria.org/research-programs/former-projects/assessm ent-of-risks-in-the-niger-republic-and-on-the-niger-nigeria-border (accessed February 2019).

Imobighe, T. A. 2001. An Overview of the Theoretical Issues in African Security. In R. A. Akindele and B. E. Ate (Eds), *Beyond Conflict Resolution: Managing African Security in the 21st Century*. Ibadan: Vantage Publishers, pp. 39–56.

International Court of Justice (ICJ) 2002. International Court of Justice gives Judgment in Cameroon-Nigeria Boundary Dispute. 10 October.www.un.org/press/en/2002/icj603.doc.htm (accessed February 2019).

International Crisis Group 2017. Fighting Boko Haram in Chad: Beyond Military Measures. Report no. 246, 8 March. Brussels, Belgium. www.crisisgroup.org/africa/central-africa/chad/246-fighting-boko-haram-chad-beyond-military-measures (accessed March 2019).

Irabor, J. 2018. Update on Security Situation in the Lake Chad Basin. Paper presented at the International Conference on Saving the Lake Chad to Revitalize the Basin's Ecosystem for Sustainable Livelihood, Security and Development, organised by the Federal Government of Nigeria and the Lake Chad Basin Commission, 26–28 February, Transcorp Hilton Hotel, 1 Aguiyi Ironsi Street, Maitama, Abuja.

Joel, D. and Akioye, S. 2017. Scores Die as Air Force Jet Bombs IDP Camp in Error. *The Nation*, 18 January. http://thenationonlineng.net/scores-die-air-force-jet-bom bs-idps-camp-error/ (accessed February 2019).

Key Informant Interview. 2018. Soldiers in Borno and Adamawa States, February.

Kindzeka, M. E. 2015. Cameroon Calls for More Vigilantes against Boko Haram. *Voice of America*, 15 December. www.voanews.com/a/cameroon-calls-for-more-vi gilantes-against-boko-haram/3103493.html (accessed February 2019).

Kowalewski, D. 1991. Counterinsurgent Vigilantism and Public Response: A Philippine Case Study. *Sociological Perspectives* 34(2): 127–144.

Lewis, J. I. 2016. How Does Ethnic Rebellion Start? *Comparative Political Studies* 50 (10): 1420–1450.

Mauro, R. nd. *Boko Haram: Nigerian Islamist Group*. Clarion Project Special Report on National Security.

Missbach, A. 2014. Doors and Fences: Controlling Indonesia's Porous Borders and Policing Asylum Seekers. *Singapore Journal of Tropical Geography* 35(2): 228–244.

Moore, M. 2001. *The Ethics of Nationalism*. Oxford: Oxford University Press.

Moore, M. 2015. *A Political Theory of Territory*. Oxford: Oxford University Press.

Musa, N. 2016. Two Vigilante Members, Several Boko Haram Terrorists Killed in Fresh Chibok Clash. *The Guardian*, 27 November.https://guardian.ng/news/two-vigilante-members-several-boko-haram-terrorists-killed-in-fresh-chibok-clash/ (accessed February 2019).

Nail, T. 2016. *Theory of the Border*. New York: Oxford University Press.

Nigeria Stability and Reconciliation Programme 2017. *Perceptions and Experiences of Children Associated with Armed Groups in Northeast Nigeria.* www.nsrp-nigeria.org/wp-content/uploads/2017/03/Research-Report-Children-Associated-with-Armed-Groups.pdf (accessed March 2019).

Norman, K. P. 2016. Between Europe and Africa: Morocco as a Country of Immigration. *The Journal of the Middle East and Africa*, 7(4): 421–439.

Nwozor, A. 2013. A Reconsideration of Force Theory in Nigeria's Security Architecture. *Conflict Trends* 2013(1) (January): 18–25.

Okolie-Osemene, J. 2016. Countering Terrorism: Interrogating Communication Oversight. In N. Maheshwari and V. V. Kumar (Eds), *Military Psychology: Concepts, Trends and Interventions*. New Delhi: Sage Publications, pp. 298–316.

Okolie-Osemene, J. 2017. Countering Insecurities in Daily Life: A Study of Vigilantes around the Lake Chad Basin. Paper presented at International Conference on Insecurities in the Lake Chad Basin, hosted by the Mega Chad Network in Université Nice-Sophia Antipolis, Nice, France, 14–16 June.

Okolie-Osemene, J. 2018. Towards Enhancing the Collaboration of State Security Providers for Safer and Peaceful Lake Chad Basin. Paper presented at the International Conference on Saving the Lake Chad to revitalize the Basin's Ecosystem for Sustainable Livelihood, Security and Development, organised by the Federal Government of Nigeria and the Lake Chad Basin Commission, 26–28 February, at Transcorp Hilton Hotel, 1 Aguiyi Ironsi Street, Maitama, Abuja.

Okolie-Osemene, J., and Okolie-Osemene, R. I. 2017. The Challenges and Prospects of Security Sector Manoeuvrability over Terrorism in Somalia. In S. Romaniuk, F. Grice, D. Irrera and S. Webb (Eds), *The Palgrave Handbook of Global Counterterrorism Policy*. London: Palgrave Macmillan, pp. 925–943.

Olaniyan, A. 2017. Foliage and Violence: Interrogating Forests as a Security Threat in Nigeria. *African Security Review* 22(1): 1–20.

Olanrewaju, T. 2017. Nigeria, Cameroon Military Parley over Border Patrol. *The Sun*, 11 July. http://sunnewsonline.com/nigeria-cameroon-military-parley-over-border-patrol/ (accessed February 2019).

Olojo, A. E. 2014. *Muslims, Christians and Religious Violence in Nigeria. Patterns and Mapping (June 2006–May 2014)*. IFRA-Nigeria working papers series no. 33. www.ifra-nigeria.org/IMG/pdf/muslims-christians-religious-violence-nigeria.pdf (accessed February 2019).

Omorotionmwan, J. 2017. Lest we Forget, What of "Mutiny 66"? *Vanguard*, 9 November. www.vanguardngr.com/2017/11/lest-forget-mutiny-66-josef-omorotionmwan/ (accessed February 2019).

Onapajo, H. 2017. Has Nigeria Defeated Boko Haram? An Appraisal of the Counter-Terrorism Approach Under the Buhari Administration. *Strategic Analysis* 41(1): 61–73.

Oshita, O. O. 2010. Conclusion: Border Security, Regionalism and Governance in Africa. In C. O. Bassey and O. O. Oshita (Eds), *Governance and Border Security in Africa*. Lagos: Malthouse Press.

Panwar, N. 2017. Explaining Cohesion in an Insurgent Organization: The Case of the Mizo National Front. *Small Wars & Insurgencies* 28(6): 973–995.

Pérouse de Montclos, M. A. 2016. A Sectarian Jihad in Nigeria: The Case of Boko Haram. *Small Wars & Insurgencies* 27(5): 878–895.

*Premium Times*2012a. Boko Haram Has Infiltrated My Government, Says Jonathan. *Premium Times*, 8 January. www.premiumtimesng.com/news/3360-boko-haram-has-infiltrated-my-government-says-jonathan.html (accessed February 2019).

*Premium Times*2012b. Nigeria, Cameroon Sign Trans-Border Security Agreement. *Premium Times*, 28 February. www.premiumtimesng.com/foreign/3983-nigeria_cameroon_sign_trans-border_security_agreement.html (accessed February 2019).

*Premium Times*2014. Jonathan gets approval to borrow $1billion to fight Boko Haram. *Premium Times*, 25 September. www.premiumtimesng.com/news/168645-jonathan-gets-approval-to-borrow-1billion-to-fight-boko-haram.html (accessed February 2019).

*Premium Times*2017. U.S. Approves Sale of N219 Billion Attack Planes to Nigeria. *Premium Times*, 4 August. www.premiumtimesng.com/news/headlines/239198-u-s-approves-sale-n219-billion-attack-planes-nigeria.html (accessed March 2019).

Raleigh, C. and Dowd, C. 2013. Governance and Conflict in the Sahel's "Ungoverned Space". *Stability: International Journal of Security & Development* 2(2): 1–17.

Rineheart, J. 2010. Counterterrorism and Counterinsurgency. *Perspectives on Terrorism* 4(5): 31–47.

Rufai, S. A. 2017. Of Ungoverned Space, Cattle Rustling and National Security. *The Guardian*, 12 April. https://guardian.ng/opinion/of-ungoverned-space-cattle-rustling-and-national-security/ (accessed February 2019).

Rumford, C. 2006. Theorizing Borders. *European Journal of Social Theory* 9: 155–169.

Sanchez, W. A. and Illingworth, E. 2017. Can Governments Negotiate with Insurgents? The Latin American Experience. *Small Wars & Insurgencies* 28(6): 1014–1036.

Schiller, N. G. and Salazar, N. B. 2012. Regimes of Mobility Across the Globe. *Journal of Ethnic and Migration Studies* 39(2): 183–200.

Schnyder, M. 2017. No Borders: The Politics of Immigration Control and Resistance. *Global Change, Peace & Security* 29(3): 317–319.

Schnyder, M. 2017. Key Informant Interview, with the authors, 25–29 December.

Shalangwa, M. W. 2013. *The Nature and Consequences of Armed Banditry in Border Communities of Adamawa state, Nigeria*. A thesis submitted to the school of postgraduate studies, Department of Sociology, Ahmadu Bello University, Zaria, in partial fulfilment of the requirement for the award of a Master Degree in Sociology.

Sharma, P. K., Ratman, A. and Madhusudhan, T. 2016. Misconduct Behaviour in Armed Forces. In Nidhi Maheshwari and Vineeth V. Kumar (Eds), *Military Psychology: Concepts, Trends and Interventions*. New Delhi: Sage Publications, pp. 119–148.

Stevenson, J. 2015. *Statistical Analysis of Event Data Concerning Boko Haram in Nigeria (2009–2013)*. National Consortium for the Study of Terrorism and Responses to Terrorism a Department of Homeland Security Science and Technology Center of Excellence Based at the University of Maryland.

Sulaimon, S. 2014. How Nigerian Troops Set up Ambush for Boko Haram Leader. *The Trent Online*, 25 September. www.thetrentonline.com/nigerian-troops-set-ambush-boko-haram-leader-abubakar-shekau-killed-graphic-photos-8/ (accessed February 2019).

Sundar, N. 2010. Vigilantism, Culpability and Moral Dilemmas. *Critique of Anthropology* 30(1): 113–121.

Taylor, A. J. 2016. Thoughts on the Nature and Consequences of Ungoverned Spaces. *SAIS Review of International Affairs* 36(1): 5–15.

Ugonna, C. 2014. Jonathan Seeks to Borrow $1 Billion to Fight Boko Haram. *Premium Times*, 16 July. www.premiumtimesng.com/news/165053-jonathan-seeks-to-borrow-1-billion-to-fight-boko-haram.html (accessed February 2019).

US Government 2012. *Guide to the Analysis of Insurgency 2012.* Central Intelligence Agency. https://fas.org/irp/cia/product/insurgency.pdf (accessed March 2019).

Weeraratne, S. 2017. Theorizing the Expansion of the Boko Haram Insurgency in Nigeria. *Terrorism and Political Violence* 29(4): 610–634.

Yamamoto, D. 2013. The Growing Crisis in Africa's Sahel Region. Testimony by Acting Assistant Secretary, Bureau of African Affairs, US State Department House Subcommittees on Africa, Global Health, Global Human Rights, and International Organizations and Terrorism, Nonproliferation, and Trade and Middle East and North Africa, 21 May.

5 Border fragility and the causes of war and conflict in the Democratic Republic of the Congo

Nelson Alusala

Introduction

The occurrence of civil wars in Africa has often involved cross-border components, whether as causes or effects. One cannot therefore practically delink the causes of territorial (inter-state) wars from those of intra-state (internal/civil) wars, as even civil wars have the potential of degenerating into inter-state wars. It is for this reason that causes of conflict should be viewed in two dimensions. Firstly, as underlying (or core/root) causes and secondly as proximate causes. This chapter examines the root and proximate causes of the civil wars that occurred in Zaire (now, and hereafter, the Democratic Republic of the Congo or DRC). It uses the civil war that occurred from 1996 to 2002 to illustrate how proximate and core issues combined to trigger the war. This chapter does not aim to verify or sharpen existing theories, but rather to build on the theoretical framework for understanding how core and proximate factors have contributed to the occurrence of the civil war in the DRC. The Great Lakes Region (GLR) of Africa is a geographical area with peculiar characteristics that range from highly tenuous inter-ethnic relations with strong regional dynamics to rich natural resources that are unevenly distributed across the countries that form the region. The countries forming the GLR can be classified into two: those in the periphery and those at the core, depending on their geographical location and role in the cross-border conflicts that have characterised the region.

Three countries, the DRC, Burundi and Rwanda, are at the core of the conflict in the GLR (Mutta, 2004; Edmonds et al., 2009). The three have a common historical and social background owing to Belgian colonial rule and the subsequent internal conflicts within their respective territories, which have often traversed their borders since independence. The countries that are proximate to the three core countries include those that geographically share the waters of the two large lakes in the region (i.e., Lake Victoria and Lake Tanganyika). These include Tanzania, Kenya and Uganda. They are also member states of the International Conference on the Great Lakes Region (ICGLR). The other parties of ICGLR, which do not necessarily share lakes and/or physical borders with the core countries include Angola, Central

African Republic (CAR), Republic of Congo, Rwanda, Sudan, South Sudan and Zambia. These countries are also signatories to the Great Lakes Peace and Security Framework of 2013.

The rest of the chapter is structured as follows. First, it briefly explains the concepts borders and core and proximate causes of conflict. Next it situates the conflict in the DRC in and discusses the characteristics which are common to the core countries of the GLR, and the region's weaknesses. It then provides a typology of the core and proximate causes of conflict in the DRC. The last section concludes.

A conceptual consideration of borders and the core and proximate causes of war and conflict

This chapter argues that proximate factors played a major role in the occurrence of the civil war in the DRC or then-Zaire, from 1996. It describes how proximate factors can influence systemic conditions in a given environment (in this case the DRC), to give rise to either violent reactions or peaceful ways of dealing with conflicting interests. Proximate factors can therefore play either a positive or negative role in a conflict.[1] "Proximity" in this case relates to the relationship of issues and the conflict at hand (i.e., how the issues impact on the conflict), and not necessarily physical location. In some cases, however, proximate causes tend to (coincidentally) traverse (or flow into) neighbouring countries due to historical and/or structural factors. In the case of the civil war in the DRC, for instance, this chapter shows that the proximate causes were facilitated by not only the porosity of the border between the DRC and Rwanda, which allowed the physical movement of refugees across the border, but also by the historical occurrences and linkages between the people inhabiting the two countries.

Although the physical border played a major role in the civil war that started in the DRC in 1996, it was not a cause. The border in this sense, as Van Houtum (2012: 674) emphasises, is not ontological (i.e., physical), but rather epistemological (i.e., issues-based) in relation to the spatial spread of occurrences. The conception is that the causes of conflicts do not necessarily require a shared physical border for them to influence a given conflict. Agnew (2008: 177) argues that the emphasis should be on how borders are constructed socially rather than simply taking their existence for granted. Agnew goes on to argue that, previously, "forces" and "functions" defined borders, but now, "discourses" and "practices" are at play. He reflects on the spatial nature of borders with respect to social life. He observes that the more recent literature on borders focuses on borders as socio-territorial constructs that reflect the discourses and practices of national identity, and concludes that the emphasis is on how they are constructed socially rather than simply taking their existence for granted. This, he states, reflects a completely territorialised image of spatiality. What this means is that borders are becoming more issue-based than the physical boundaries between states. Therefore, when talking of proximate causes of conflict with respect to the DRC and its neighbours, the concept of border refers to not only the physical boundary lines that divide these countries, but rather/also the social construct of the border in terms of how people live and relate with each other in the region and beyond, and how those relationships either ease or heighten the likelihood of an open conflict.

The discussion of core/root causes of conflict in this chapter follows the logic of intractable conflicts. Such conflicts, according to Coleman (2011: 428), defy attempts to resolve them, and are characterised by high levels of intensity and destructiveness. They also typically involve many actors and are built around an intricate set of historical, religious, cultural, political and economic issues. Their prolonged nature makes them structural and according to Coleman, because they are central to human social existence they tend to resist any attempts to resolve them. Their centrality to human survival makes parties reject negotiations, as each party views the rigid position of the other as a threat to its own existence. Ultimately, the prevailing feeling of insecurity leads to persistent threats and hostilities that characterise the everyday lives of the parties concerned, and supersedes their ability to capitalise on any shared concerns they might have. This is anchored in the human needs theories, which argue that some of the core causes of conflicts are related to a failure by society to provide fundamental human needs (Maslow, 1973). These include not only the basic needs for food, water and shelter, but also more complex needs for safety, security, self-esteem and personal fulfilment (Burton, 1990). These more complex needs centre on the capacity to exercise choice in all aspects of human life and to have one's identity and cultural values accepted as legitimate. Moller (2003: 10) examined these two categories of causes of civil war to reach the conclusion that conflicts bear both structural (core/root) causes as well as proximate (trigger) causes. According to Moller, conflicts do not reach a full-fledged level without the combination of core and proximate causes. In other words, a single event cannot lead to an open conflict unless there are underlying structural reasons. As elaborated in this chapter, some of these needs constitute the core causes of the problems that the DRC continues to face.

Locating the DRC, and war and conflict in the DRC, in the Great Lakes Region

The three core countries in the GLR conflict share common characteristics, including a common colonial past, ethnic identity, internecine conflicts, regime weakness and economic decay, as discussed below.

A common colonial past

Walter (1998), Prunier (1995), Reyntjens (1996) and Collins (2008) point to the existence of common characteristics between Rwanda, Burundi and the DRC. The most disturbing common reality among them is their conflict-laden instability both in the past and in the present. The three countries have known little peace and tranquillity since their respective independences in the 1960s. Other common attributes of the three states are that they share the common colonial experience of having been ruled by Belgium and that they have been scenes of vicious ethnic conflicts. There is chronic disaffiliation of substantial proportions of their populations from national political structures. In addition, the three are contiguous, with ethnic associations transcending the political frontiers (Walter, 1998: 75–76). This makes it easy for any conflict to spread across borders. These are core issues that define the way the three countries relate with each other.

Part of the blame for the three countries' intractable conflicts has been attributed to Belgium, the former colonial power. Belgium is blamed for withdrawing hurriedly and recklessly in 1960, leaving behind three unprepared, newly independent states. In all the three states (though in the case of the DRC this applied largely to its eastern provinces) the Belgians governed in an alliance principally with and through one minority ethnic group, the Tutsi, whose approach to power, according to Edmonds et al. (2009: 8), was different from that of Belgium. The departure of Belgium therefore created a sudden void that led to ethnic power struggle.

A common ethnic identity?

Ethnicity is one of the major drivers of the conflict in the three countries. The origin of the ethnic identity crisis of the 1990s in Rwanda and Burundi goes back to the 1950s. The two share closer common demographics in which a predominant ethnic group, the Hutu (or *Bahutu*), comprise about 85 per cent of the population, followed by the Tutsi (or *Batutsi*) with about 14 per cent of the population, and the *Twa* (or *Batwa*), a pygmy forest group comprising less than one per cent of the total population (Prunier, 1995: 389). On the other hand, approximately half of the populations of the Kivu Province of Eastern DRC owe their origins to Hutu and Tutsi family groups who immigrated from spaces that are the present Rwanda and Burundi over the past 400 years, when the borders as we know them today, were nonexistent.

In 1959, the Belgian trusteeship administration supported a Hutu revolution in Rwanda, which led immediately to widespread killing of Tutsi and many refugees fleeing into Uganda and other surrounding states (Reyntjens, 1996: 240–251; Walter, 1998: 78–79). By the 1980s, some 480,000 Rwandan Tutsi had taken refuge in neighbouring countries, and 240,000 Hutus had fled from Burundi by 1991 (Lemarchand, 1996: 63). In describing the evolution of events in Burundi, Chrétien notes that about four decades after independence, the situation in Burundi had completely changed, to resemble Rwanda's Hutu/Tutsi cleavage that had, from historical times permeated the central logic of political and military arrangements, engendering violence, fear and hate (Chrétien and Banégas, 2008: 23). He adds that the assassination of President Ndadaye of Burundi in 1993 marked the failure of the democratic experiment in the country, and the heightening of ethnic tensions that later exploded in the GLR as a whole.

Harff and Gurr (2004) promote the view that social classes and distribution of resources are the sole issues at the heart of international relations, affirming that the state remains the main actor in international relations. It should, therefore, be better thought of as a recognised territorial entity in flux and not a monolithic enterprise. This is because ethnic groups forming a state are rarely equal in power, legitimacy or with regard to how they access economic resources. Harrf and Gurr go further to question whether the institutionalised state has a future, given the many ethnic groups that clamour for independence. They concur that since ethnic groups are in search of the right to govern their own territory, which they hope will become a sovereign, internationally recognised state, the current international system may fragment into hundreds of mini-states unless ethnic demands can be satisfied within existing states. Similarly, Sambanis (2003) and Kalron (2010) argue that ethnicity is a major factor in the occurrence of civil wars, as exemplified by the civil wars in the GLR. The core of these wars, argues Kalron (2010: 25–37) has its base in the activities of ethnic groupings inhabiting the DRC, Rwanda, Burundi and Uganda. Kalron affirms that ethnic sentiments in the region go back to pre-colonial times. The web of relationships is both by diffusion and contagion, as elaborated under proximate causes of civil war, later.

As the conflict in Rwanda escalated, there was a breakout of ethnic violence in Burundi, followed by the assassination, in October 1993, of Malchior Ndadaye, Burundi's first elected Hutu President. These events triggered the migration of Burundian Hutu refugees into Zaire's South Kivu Province (Evans, 1997). The unregulated border and the pre-existing cross-border family ties made the spread of the conflict into Zaire unavoidable. The entry of Hutu refugees from Rwanda and Burundi put them in collision with the *Banyamulenge*, an ethnic group of Congolese Tutsis inhabiting Mulenge Mountains in South Kivu Province, which considered Hutus as a threat to their domination in South Kivu, especially with regard to land and access to other exploitable natural resources. This escalated pre-existing ethnic tensions in Eastern Zaire. The consequent outbreak of the genocide in Rwanda in 1994 exacerbated the situation with the arrival of *génocidaires* from Rwanda.

The relationship between the civil war that occurred in Rwanda in the early 1990s and the civil war that broke out in Zaire in 1996 is a linear one. The latter was a continuation of the former, having spread across the porous border. A large number of high-ranking military officers, as well as thousands of heavily armed *Interahamwe* (the Hutu militia at the forefront of the genocide in Rwanda) and the majority of the Rwandan forces (now known as ex-FAR) managed to escape the inexorable advance of the Rwandan Patriotic Front (RPF) by retreating to safe zones.

Once it was clear the RPF could not be halted, France facilitated the escape of much of the Hutu power leadership into Zaire (Bonner, 1994). This action in itself had a proximate impact on the already fragile internal structures of Zaire. Again, the porosity of the borders between the two countries made such escape easy. The resultant spillover and entwinement of national conflicts – especially between Rwanda, Burundi and the DRC, created a spatial web of ethno-political conflict within and across the borders of these countries, leaving an indelible mark on the peoples of the region, characterised not only by a mix of ethnic linkages, but also similar conflict experiences.

The cyclic occurrence of civil wars in Rwanda and Burundi from historical times continued to produce refugees in the region, with most of them entering the vast and fertile Eastern Zaire which, after several centuries, they called home. However, the fact that these *Kinyarwanda*-speaking communities (or *Rwandaphones*) maintained their language and ethnic affiliations to Rwanda, demonstrates how language, spatially distributed through movement of people, can become an identity issue, especially in terms of conflict. The Rwandaphones encountered gradual resistance from other Zairian tribes such as the Hunde, Nyanga and others, who considered them "outsiders". This was partly the driving force behind the RPF resolve to overthrow the Hutu establishment in Rwanda, so as to pave way for the return of exiled and alienated Tutsi from neighbouring countries and the diaspora back to Rwanda (Chrétien and Banégas, 2008: 207). The spatial inter-relationship among various ethnic communities in Rwanda, the DRC and Burundi is such that one cannot discuss the issue devoid of linking it to the neighbouring states. Lake and Rothchild (1996: 41–75) observe that in the three countries, the ethnic card has been played over time by politicians with the aim of instigating inter-ethnic violence and to mobilise support. As such, every fresh episode of Hutu–Tutsi violence in Rwanda or Burundi resulted in refugees fleeing to neighbouring countries, especially into Eastern DRC.

The genocide that followed in Rwanda and the killings in Burundi exceeded the mortality rate of any other conflict (except Cambodia) since the Second World War. McGreal (1996: 11) records that in Rwanda, an estimated 500,000–800,000 deaths occurred in 1994, while in Burundi the accumulated death toll in successive violent episodes between 1993 and 1997 was at least 150,000 people. This depicts the extent to which ethnicity can be manipulated into a core/root cause of a conflict.

The entry of Rwandan refugees into the DRC, accompanied by *génocidaires* fleeing from advancing RPF troops exported the conflict into the DRC, although the core issues of the conflict were spatially present within the communities inhabiting the region. The historical cross-border ethnic linkages and endemic conflicts facilitated the ease with which the effects of the civil war in Rwanda diffused into the DRC, which was also experiencing internal structural weaknesses. Such was the proximate impact that the Rwandan civil war had on the DRC, triggering the war in the DRC. The porosity of the borders in the GLR, and particularly between the DRC and Rwanda, only facilitated the merging of proximate and core causes in the DRC, thereby triggering a violent conflict.

Internecine conflicts

The endemic nature of the conflicts in the three countries of the GLR is captured by Daley (2006: 303) who writes that since 1960 there have been 11 outright wars in the DRC, five in Burundi and two in Rwanda, while low-intensity warfare has existed at all times. As such, more than four million people have lost their lives as a direct result of these conflicts, while a further four million have been rendered refugees and internally displaced persons (IDPs). The 1994 genocide in Rwanda marked a decisive moment in the history of the region's conflict, while the involvement of seven African nations in the DRC war of 1998–2004 (also referred to as Africa's First World War), escalated what had started in Rwanda as a civil war, into a regional conflict. In his discourse on war, violence and political re-composition in Africa, Banégas (2008: 59–79) contends that the African GLR has been the theatre of traumas that have adversely disrupted the geographical, economic, social and demographic stability of the region, ranging from the civil war in Rwanda in the 1990s and the massacres that followed; the assassination of President Ndadaye of Burundi in 1993; the two Congo wars of 1996 and 1998 that led to the fall of President Mobutu Sese Seko and, later, President Laurent Kabila, and the fragmentation of the DRC; the proliferation of rebellions and armed movements within the DRC; and the installation by armed men, of economies of predation throughout the region.

Regime weakness and economic interests

These two have been the major contributors to the continuous warfare in the GLR. The consequences of the fragility of nation states in the region exhibited from the early 1990s coincided with the end of the Cold War, which exposed the strife which was previously concealed by Cold War politics. This became more evident as a result of the unravelling of aspects such as the authoritarian political regimes, the economic decay of the 1980s, and the renewed communal contests for power and resources.

Ewald (2004: 26) argues that:

> unlike previous periods, these conflicts became regionalised, transforming the Great Lakes into a trans-African belt of conflict and discord. The collapse of the predatory Idi Amin regime in Uganda in 1979 set the pace for the convulsions that were to engulf the region through contagion, subsequently embroiling the equally weak, but authoritarian governments in Rwanda and Zaire.

Ewald concludes that in Uganda, the changes were instructive to Rwanda's Hutu–Tutsi divide, when Yoweri Museveni's National Resistance Movement (NRM) restored stability after a devastating conflict. The large population of Rwanda's Tutsi refugees that had survived the Hutu Revolution of 1959 quickly copied the NRM experience. The rise of Museveni from 1986 produced two outcomes of significant consequence to the region. Firstly, it rejuvenated the national institutions in Uganda, and secondly it led to the resurgence of the national question between Hutu and Tutsi that quickly spread within and outside Uganda's borders, thereby sharpening ethnic identities in Rwanda and Burundi. This situation played a significant role in setting the platform upon which the events from 1990 to the first Congo War (1996–1997) and the second Congo War (1998–2004) sprung. Three major factors helped to weaken the political regimes in the three countries.

Firstly, the colonial period, particularly that under Belgium, instrumentalised identity issues, thereby putting at loggerheads the Hutu and Tutsi groups that shared the same language, culture, history, social organisation and territory. Secondly, Mobutu's long patrimonial rule and the manipulation of ethnic identities sharpened the already sensitive issue of ethnicity, especially in the eastern region of Zaire. Thirdly, the effects of the civil wars that occurred in Burundi and Rwanda spilt into Eastern Zaire, whose governance structures were weak. The already illegal exploitation of natural resources that was ongoing in the east of Zaire attracted belligerents from the neighbouring countries, who saw an easy opportunity for building an economic base to finance their wars back home. This transformed Zaire's internal conflict into a regional war, a discussion that lies beyond the scope of this chapter.

Economic decay

A combination of internecine violence and weak and corrupt regimes have led to high levels of poverty in the three core countries of the GLR. According to Hochschild (1998: 32–40), the economies of these countries are dependent on resource extraction under conditions of endemic structural violence, and as such they continue to record low economic growth, a situation that Harvey (2004) refers to as "accumulation by dispossession". This is reflected in the troubled nature of politics from independence to present – characterised by genocide, forced labour and displacement – making their incorporation in the global economy difficult. The masses in these countries have reaped no benefits from the vast natural resources that their countries own, leading to low life expectancy and disruption of livelihood strategies.

The situation in these countries fits into what theorists of peace such as Galtung (1969: 167–191) refer to as structural violence – an endemic situation in which poverty and powerlessness constitute an indirect form of violence, leading to outright violence, as in civil wars. Positive peace can only result if these conditions are reversed and social justice allowed to prevail.

Other scholars of the GLR have attributed the Hutu–Tutsi conflict to various causes, ranging from environmental, to ethnicity and population density. Uvin (1998) argues that Rwanda's Hutu–Tutsi rivalry, whose effects have spread into the DRC and Burundi, stems from frustrations inherited from a socially and culturally oppressive environment that goes back to the colonial administration, and is reinforced by the genocidal ideology developed after the revolution of 1959–1961. Marysse, De Herdt and Ndayambaje (1994) and Willame (1995) on the other hand, posit that the economy of Rwanda has deteriorated steadily especially from the 1980s, thereby sharpening further inter-group inequalities that underlie the rivalry between the two groups that spread in the region. They add that the country's high population density has also contributed to inter-group tensions.

Reyntjens (2006: 15–18) makes the case that there are multiple causal factors that explain the ethno-political conflict in the region. He argues that democratisation and control of state are proximate causes common to these three countries as well as other African countries. The state in Africa is therefore an important instrument of wealth accumulation and reproduction of a ruling elite. As such, in most cases political transitions involve violence (Huntington, 1993). Kabanda (2008) reinforces Reyntjens' discourse on ethnicity by adding that the epicentre of the GLR conflict lies in the inter-group differences between Hutu and Tutsi in Rwanda and Burundi. He maintains that although the circumstances in each country determine the course the conflict takes, the civil wars that have occurred in Rwanda and Burundi, and spread into the DRC, stemmed from the historical hatred for the "other". Kabanda links the Rwandan scenario to the Burundian one by arguing that the phenomenon of hatred for the "other" goes back to 1959 in Rwanda, the time when the Burundian party's ideology was imported from the Rwandan *Palipehutu*. Since then, the violence in Burundi has been either as a result of subsequent efforts to import the Rwandan model or as a fearful reaction to it. He concludes that the Hutu–Tutsi conflict in the Kivus in Eastern DRC is attributable to the ideological contagion emanating from the 1959 Rwandan revolution. It is within the broader context of the GLR discussed above that one needs to understand and disaggregate the causes of the war in the DRC on the basis of core and proximate issues.

Typology of the core and proximate causes of the war in the DRC

Core or root causes of the civil war that has endemically affected the DRC revolves around the country's natural resources and regime-type. They may manifest themselves through poverty, political repression and (ethnic) marginalisation, prolonged failure of state functions and unequal distribution of resource benefits. A combination of these factors, according to the UN Secretary General's Millennium Report, provides the seedbed of internal conflict (United Nations, 2000: 44–45). The report also underscores that the majority of wars today are due to competition for dwindling resources. Primarily, poverty, coupled with sharp ethnic or religious cleavages lead to stiff inter-group competition and therefore conflict. The situation is further worsened by insufficient inclusiveness of government institutions and where the allocation of society's resources favours one dominant faction over others.

A close analysis of the history of (civil) wars in the DRC demonstrates how the core and proximate causes of war influenced the conflict in the country. Proximate issues that influenced the internal state of affairs in the DRC included changes in the strategic environment brought about by the Cold War; the global democratisation demands that emerged in early 1990s and which called for multi-party politics; and the use of ethnicity as a political tool. These factors, exacerbated by the civil war and genocide in the neighbouring Rwanda, merged core and proximate issues, leading to the war in the DRC. As the Rwandan Patriotic Army (RPA) rebels advanced into Rwanda from Uganda, they defeated the Rwandan government forces, that is, the Rwandan Armed forces or *Forces Armées Rwandaises* (FAR). As the RPA defeated the FAR, they pushed them and thousands of refugees into Eastern DRC. This, and the conflict that emerged in Burundi in 1993, had a contagion effect on war in the DRC. The simultaneous emergence of conflicts in the three countries of Rwanda, Burundi and the DRC in the early 1990s put the three at the core of the conflict that spread into the DRC. The ontology of the cross-border linkages amongst the people of the three countries can best be appreciated when viewed in a regional context.

Core or root causes

The civil wars in the DRC were influenced by changes in the strategic environment and regime change, global democratisation demands as well as the role of ethnicity as a political tool.

Demands for democratisation

Faced with the domestic and foreign political pressure of the early 1990s, coupled with a rapid decline in the DRC's domestic economy, President Mobutu reluctantly moved towards democratisation and liberalisation. In April 1990, the prevailing conditions compelled him to make political concessions. He legalised independent opposition political parties, thereby putting in place a multi-party democracy (Clark, 1998: 25–32). A national multi-party conference (Sovereign National Conference – SNC) was convened in August 1991 with the objective of drafting a new constitution and preparing for elections. However, the conference turned chaotic and disintegrated after only a week due to ethnic and political rivalries. A month later Mobutu agreed to form a coalition government with Union pour la Démocratie et le Progrès Social (UDPS) leader Etienne Tshisekedi (Mobutu's opponent since 1981), as Prime Minister. However, Tshisekedi was later sacked after only month in office (October 1991). Another attempt at holding the conference in January 1992 failed when Mobutu's handpicked Prime Minister, Nguza Karl-I-Bond dissolved it unexpectedly (Clark, 2007: 44–45). The mandate of the Conference was to chart a new (multi-party) structure. Eventually the SNC was convened in August 1991, and dragged on for 18 months due to

Mobutu's reluctance to share power. In August 1992, the SNC drafted the Transition Charter (La Chatre de la Transition) to replace the constitution.

According to the Charter, Mobutu was supposed to share certain ministerial positions with the opposition, such as finance, foreign affairs and defence. In addition, the Conference had voted to reinstate Tshisekedi to the position of Prime Minister. The struggle for power took an ethnic angle when the Lunda tribe of Shaba (supporters of Nguza), clashed with the Luba of Kasai (Tshisekedi's tribe). Each ethnic group mobilised against the other and soon the Shaba announced that they would not recognise Tshisekedi's authority. This resuscitated the memories of Katangese session claims of the 1960s when they demanded for independence from Kinshasa. Clark (1995: 25–30) writes that at the same time as the Luba of Shaba were being deported to Kasai from Shaba where they had lived for ages and their property sequestered, in North Kivu the Tutsi were being attacked by the autochthons due to ethno regional contestation as a result of the spillover effect of the Rwandan civil war and the genocide.

Mobutu's final attempt to salvage the collapse of Zaire came in 1994 when he named Kengo wa Ndondo as Prime Minister. As this was happening, the contagious effect of thousands of refugees fleeing the genocide in Rwanda started to be felt in the Kivus. In a move meant to salvage his political limelight, Mobutu instituted a policy to expel the Banyamulenge, as a move to avert the potential danger of the Banyamulenge offering support to the RPF in revenge for the support that Mobutu had provided President Habyarimana. The Banyamulenge's concerns are linked to Rwanda's invasion of the DRC in 1996. However, the ethnic tensions that surrounded Mobutu's decision to expel the Banyamulenge comprise the very core of ethno-politics that characterised the conflict, as discussed below.

Ethnicity as a political tool

Although the war in the DRC is directly linked to the Rwandan civil war, the latter could be best described as a proximate cause. The ethnic communities inhabiting the North and South Kivu Provinces of Eastern DRC, like their counterparts in Rwanda and Burundi, have experienced interethnic conflicts from historical times. Long before the Rwandan civil war of 1990 broke out, ethnic conflicts were already raging in (Eastern) Zaire. For instance, in 1989 Mobutu's regime yielded to the historical pressure to undertake an exercise aimed at identifying "foreigners" among the Congolese in North and South Kivu. The Rwandaphones in North Kivu are referred to as Banyarwanda, while in the South Kivu they are called Banyamulenge. The way in which the question of nationality was carried out under the regime of Mobutu set a foundation for ethnic differentiation in Zaire. The process of identifying "foreigners", which was known as *Mission d'Identification des Zairois au Kivu* (MIZK), entailed carrying out a physical verification of the nationality rights of all Kinyarwanda-speakers as a way of authenticating the "reality" (Koen, 2002: 8–9).

The mission, which started in 1991, was manipulated by local members of the smaller autochthonous ethnic groups in an attempt to disenfranchise the Banyarwanda and Banyamulenge of their accession to national citizenship (Koen, 2002: 506). According to the Rwandaphones, the denial of citizenship meant Kinshasa's abrogation of the nationality law of 1981, which, according to them, had bestowed on them full political rights because they had settled in Zaire before 1885 (Koen, 2000; Ruhimbika, 2001; Van Hoyweghen and Koen, 2000). In retrospect, the autochthons insisted that the Banyamulenge could not claim any citizenship rights because in 1885 there was no ethnic community called "Banyamulenge" living in Zaire. This was an unfortunate turn of events for the Banyamulenge, a name they had adopted in the 1960s for two reasons. Firstly, they aimed to shade off the name "Banyarwanda" which had a "Rwanda" connotation. Secondly, they sought for a name that would give them solely a Congolese image. In this case, "Banyamulenge" suited them best – "those coming from the Mulenge Mountains of Eastern Zaire" because it differentiated them from the more recent Tutsi arrivals in the Kivus, particularly during the 1959 civil war in Rwanda. This was in an effort by the Banyamulenge to claim their local social and political rights. As Koen concludes:

> The term in use before (Banyarwanda) connected them with Rwanda and expressed the foreign nature of their origins. The term Banyamulenge, on the contrary, linked them to the place where they first settled and offered them an indigenous identity. Yet, this attempt to distance themselves from Rwanda was viewed by the rest of the population as an attempt to camouflage their real identity.
>
> (Koen, 2002: 501)

When the policy of MIZK failed to resolve the issue of Banyamulenge citizenship, the disgruntled youth among them sought an alternative to fight for their marginalised position in South Kivu. This was the time when the RPF recruitment and insurgency against the regime of Habyarimana was intensifying. The apathy created by abrogation of their nationality drove some of the Banyamulenge youth to sympathise with the situation of their Tutsi RPF brothers in Rwanda, thereby opting to join the RPF as early as 1991 (Koen, 2002: 508). This recruitment increased after the Rwandan genocide and the subsequent refugee crisis as the RPF prepared to attack the ex-FAR and Interahamwe in refugee camps in Zaire, after the latter declared the "Government of Rwanda" in exile in 1995. From these old rivalries emerged the ethnic factor that characterised the two Congolese wars (1996–1997 and 1998–2004).

The spread of the Rwandan civil war (between the Hutu and Tutsi) into Zaire brought to the fore the historical hatred of these two Rwandaphone groups by other Congolese tribes. The understanding of this historical relationship is central to understanding the diffusion and contagion of the

Rwandan civil war into Zaire. Without going into the ethnology of Congolese tribes, it suffices to point out that the history of the Banyarwanda in the Kivus is a complex one. Several groups of Banyarwanda people arrived at different times in history. The oldest group of the Banyarwanda families, both Hutu and Tutsi, had been living in the DRC for centuries before the arrival of Europeans. This was followed by a later arrival of mostly Hutu from Rwanda, brought in by the colonial authorities in the 1920s and 1930s to supply labour for the under populated Belgian Congo. Willame (1997: 39) points out that even though the Kivus (North and South), which is a region about four times the size of Rwanda, is less densely populated than Rwanda, it is far from empty due to its close links by demographic and political patterns, to the situation in Rwanda.

In the North Kivu Province, the local population is composed of the autochthons (*Bahunde, Banyanga, Banande*, etc.) and Kinyarwanda speakers divided up into smaller communities that included the Bahutu, Batutsi as well as Banyabwisha and Bafumbira (originally from Uganda). The situation was calmer until in 1937 when the *Mission d'Immigration des Banyarwanda* (MIB), created by Belgians, brought agricultural workers from Rwanda into the Kivus. It is estimated that during its 18 years of existence, MIB brought about 85 thousand Banyarwanda from Rwanda mostly into North Kivu, although some were sent to Katanga to work in the mines (Pourtier, 1996: 20). These were followed by a more recent group of immigrants, exclusively Tutsi, who had fled the 1959–1963 ethnic hostilities in Rwanda. Together, the North Kivu Banyarwanda formed a tightly knit, mutually supportive community without distinction between Tutsi and Hutu. They preferred to settle in Masisi and Walikale zones. Today they are said to represent about 40 per cent of the North Kivu province.

The thriving nature of the economy of Banyarwanda community in the Kivus, coupled with their feeling of exclusivity, created resentment among autochthons, especially with smaller tribes such as the Hunde and Nyanga, who already felt marginalised by the majority Nande. Pottier and Fairhead (1991: 444) argue that the Nande were less threatened by Banyarwanda because they controlled a fair quantity of the territory due to their big numbers. The general perception in the Kivus has therefore been that the presence of Banyarwanda was partly as a result of Belgian colonial policies hence the Banyarwanda landholding concepts did not get the approval of the locals. The social transformation introduced by colonial administration, in which land was buyable, clashed with the local lineage of chiefs over land distribution. The Banyarwanda were viewed as taking advantage to turn their "acquired customary land rights" into permanent ownership. Between 1960 and 1965 a civil war broke out between the Nande and Banyarwanda (popularly referred to as the Kanyarwanda Rebellion) in South Kivu, in which the local administrators sought to persecute the Banyarwanda and reclaim their land. Within the same period back in Rwanda, the Hutu–Tutsi ethnic wars continued to force more people into Zaire as refugees. For instance, the Tutsi massacres in Rwanda between December 1963 and January 1964 led to fresh arrivals into the Kivus, worsening the situation through contagion (Keane, 1995: 193–198).

The situation was such that whereas autochthon tribes had local (customary) power bases commonly referred to as *chefferies* or chiefdoms, Rwandaphones could not get a local power base except from the central government. Mobutu capitalised on this weakness to gain the support of the Banyarwanda at a time when his regime was faced with the challenges on his leadership by naming one of them, Barthelemy Bisengimana, a Tutsi refugee from Rwanda, as *chef du bureau* of the presidential office. With Bisengimana's support, Tutsis went into lucrative businesses and acquired more land, especially that abandoned by Belgian farmers or forcefully taken from them in the 1973 "Zairianisation" exercise. It is recorded that the land grabbing by Banyarwanda reached such an alarming proportion that in 1980, the Ministry of Lands in Kinshasa had to cancel an allocation of 575,000 acres to the notorious Cyprien Rwakabuba, a Mnyarwanda (Willame, 1997: 44).

In addition to the land issue, was the question of citizenship for the Rwandaphones. When Bisengimana rose to power, he influenced the political bureau of the Movement Populaire de la Révolution (MPR), Mobutu's political party (the only one that had existed since he took over power) to issue a decree on citizenship that declared persons of Rwandan and Burundian origin and residing on the Belgian Congolese territory in or before January 1950 automatically Zairian citizens. When Bisengimana lost power in 1977, the law was reversed, and a new one was passed in June 1981, abrogating the famous Article 15 that Bisengimana had invoked while declaring the citizenship of Rwandans and Burundians living in Zaire. The ensuing confusion was such that the 1987 elections could not be held in North Kivu because there was no clear definition of who was a Congolese and who wasn't, in order to draw a poll list. This coincided with the period when Mobutu's regime was faced with internal opposition to his leadership and a waning Western support in the face of the end of the Cold War.

Faced with increased political pressure, Mobutu called the Sovereign National Conference (Conférence Nationale Soveraine – CNC) in 1991. One of the results of the conference was the exposure of the tensions in the country resulting from ethnicity and social identity. At the conference, the group of autochthons from North and South Kivu sought to rescind the citizenship of Banyarwanda under the 1981 citizenship act, and force them to return to Rwanda and Burundi. At the CNC, the autochthons managed to lobby for the barring of Banyarwanda from taking their seats at the Assembly under the pretext that they were not Zairians. They further invoked the CNC decisions to overhaul the local administration in North Kivu by installing new judges and police from the Hunde, Nande and Nyanga communities (Prunier, 1995: 195). As this was happening in the Kivus, in Rwanda a civil war that had broken out between Hutu and Tutsi when the RPF invaded Rwanda from Uganda in 1990, worsened. However, it suffices to mention that following the RPF attacks, President Habyarimana and his followers organised a pro-Kigali syndicate under the cover of a peasant network known as the *Mutuelle des Agriculteurs de Virunga* (MAGRIVI) to recruit young Hutu fighters (Chajmowicz, 1996: 115–120). This had a contagious effect on the activities of the RPF, who, in return also embarked on recruiting young Zairian Tutsi into its force, as mentioned earlier. This Hutu–Tutsi ethnic-focused recruitment in eastern DRC drove a wedge between the two ethnic Rwandaphone groups (Banyamulenge and Banyarwanda), thereby weakening their unity in Zaire, at a time when they both faced an assault from the other ethnic groups in the Kivu, who viewed the Rwandaphones as intruders (Prunier, 2009: 50).

With Eastern Zaire and Rwanda gripped in the above-cited tensions, the spread of the Rwandan civil war into Zaire became inevitable. In Zaire, a low intensity conflict broke out in 1992 between Hutu and Hunde. The war soon escalated when Hunde militias in Ntoto market near Walikale killed a number of Hutus in March 1993. This was soon followed by the killing of about 20,000 Banyarwanda by August 1993 and displacing of about 250,000 people from different tribes (Tegera, 1995: 395–402). In response to this, Mobutu came to North Kivu to oversee the deployment of his famed elite *Division Spéciale Présidentielle* (DSP) in 1993, leading to an abatement of the violence and opening room for negotiations in November amongst local leaders. By February 1994, some semblance of peace had been restored in the Kivus, only to be reversed by the arrival of an estimated 850,000 Hutu refugees from Rwanda in the aftermath of the 1994 genocide. From the foregoing discussion, it emerges that long before the Rwandan civil war of 1990 broke out, ethnic conflicts were already raging in (Eastern) Zaire.

The contentious issue of the Banyamulenge nationality reached a climax at the height of the Rwandan refugee flows into Zaire in 1996, when the Governor of South Kivu at the time sought to expel them out of Zaire due to their (perceived) complicity with and sympathy for the RPF. In his analysis of the situation in Eastern Zaire at the height of ex-FAR activities Walter points out that:

> The already desperate refugee situation in Zaire became supercharged on 7 October 1996, when the deputy governor for the Southern Kivu declared that the substantial (300,000) native Banyarwanda population of the Kivu would be expelled forthwith from the country, or be "hunted down as rebels".
>
> (Walter, 1998: 9)

The above statement reinforced the Banyamulenge adherence to RPF, whom they saw as the only possible saviour of their plight in Zaire – if the RPF could conquer Zaire and replace the Mobutu regime, then perhaps their nationality issue would be solved. However, the events that followed did not facilitate the solving of the nationality question for the Banyamulenge, which remains a pernicious and divisive element of the Congolese politics.

The policy to expel Banyamulenge worsened the situation by sharpening and spreading ethnic animosity in the entire east of the DRC (Clark, 2007: 46). By July 1994, at the height of the arrival of Rwandan refugees into Zaire, Mobutu was still grappling with increasing defragmentation of his authority and the mounting resistance from Banyamulenge over the quest for their Congolese nationality. There was already an emerging power vacuum in Zaire, and, as Clark puts it, Zaire "had a hard outer layer but a fragile inner core; the core of support for Mobutu had long eroded. The periphery, although not having total autonomy from the centre, was not effectively under its control either" (Clark, 2007: 46). Mobutu's political manoeuvres had almost reached an end, and his regime had weakened considerably. The arrival of Hutu *génocidaires* including Interahamwe and ex-FAR into eastern Zaire added more tension to the already tense environment in eastern Zaire.

From the foregoing, it is conclusive that the inter-ethnic conflicts that characterised Zaire during the 42 years of the Mobutu regime helped to define his kind of leadership; one which was characterised by divisiveness, patriotism, cronyism, economic and political crises, and exclusion of ethnic groups, especially in the eastern region of the Congo (Baregu, 2006: 60). Mobutu's kind of leadership was therefore to a large extent one of the main core causes of the instability that has affected the DRC, particularly the eastern region. The political decay that marked his leadership led to the rise of political movements that sought to "liberate" the country. However, the military and political backing from Western countries during the Cold War sustained Mobutu's reign and protected him from several armed insurgencies that tried to topple him. These insurgent movements included Laurent Désiré-Kabila, under his *Partie de la Révolution Populaire* (PRP). During the first Zaire war, Rwanda and Uganda picked on Kabila (as proxy) to lead the AFDL so as to give the movement a Congolese face. The planners of the rebellion chose to start it in South Kivu by preying on the plight of the Banyamulenge's citizenship issue as a wider concern for the Kinyawanda speakers in the GLR. In this discussion, the regional complexity of ethnicity and its cross-border effect reveals itself. The aggression with which other Congolese tribes relate with the Rwandaphones exhibit mistrust and betrayal, aspects that are better understood by looking at how the civil war in Rwanda turned out to be a proximate cause of the war in the DRC.

Proximate/external causes

The first DRC war (1996–1997) was an extension of the Rwanda civil war into the DRC. When Rwanda first invaded the DRC in 1996, the reason given was to safeguard Rwanda's territorial security against the spillover effect of the war resulting from the regime change in Kinshasa. A similar reason was again used to justify the invasion of the DRC by Rwanda and Uganda in

1998, although behind this façade was, as Reyntjens (2011: 132) explains, a logic of exploitation of natural resources and a larger geographical design aimed at establishing a Rwandan space of political and military control in eastern DRC.

The actions of Rwanda and its ally Uganda in Eastern DRC put all the actors in a war-like situation, in other words, it escalated the tension especially among the former *génocidaires* and refugees inhabiting the camps located close to the Rwandan border. The actions of Rwanda and Uganda put the Interahamwe and ex-FAR militias on the defensive, triggering insurgency actions against Kigali. This ended by flaring up the conflict and entrenching it in the DRC. Although the FAR and Interahamwe militias had been partly disarmed as they entered Congo, they immediately found the urgency to re-arm due to RPF operations against them. Some of the weapons and ammunition were sold to them by the *Forces Armées Zaïroïses* (FAZ). The pains of being pursued into their hiding in the DRC developed in them the ambition of revenge. They vowed to one day claim their return to Rwanda by capturing power by all means (Rufin, 1996: 27). This, according to Rufin would entail finishing the unfinished job of genocide. Thus the first war in the DRC was a direct invasion by Rwanda and its allies (Uganda and Burundi – also referred to as the Rwandan alliance) of Zaire.

The invasion of Zaire

Under normal circumstances, the Rwandan civil war of the 1990s would have had no direct impact on the stability of DRC, even less lead to a change in DRC's political regime. It was unexpected that an extraneous occurrence of such a proximate nature would be central to the eventual collapse of the Mobutu regime and lead to cycles of wars in the DRC. The DRC wars affirm an important but often less expected lesson in wars: that although a war in one country may be expected to have only secondary effects in the neighbouring countries, such war can sometimes amplify itself into a direct (core) cause of instability in the neighbouring country. In this respect, the events in Rwanda exacerbated the civil war in the Zaire in two ways.

Firstly, the failure of the RPF to transform their tactical victory (against FAR) into strategic peace led to a protraction of the Rwandan conflict into Zaire, as the RPF pursued the FAR and Interahamwe into eastern Zaire in pursuit of total victory. Secondly, once in Zaire, the political objectives of the RPF evolved from extermination for the FAR/Interahamwe, to the overthrow of Mobutu essentially for his support of the Habyarimana regime. Furthermore, Rwanda and its ally, Uganda, also evolved economic interests (in natural resources), thereby losing focus on their original political aim. Both these factors are detailed below, including the differences upon which the Hutu–Tutsi ethno-political and religious relations are based. It is arguable therefore that the failure by the RPF to transform its tactical victory into strategic victory within Rwanda had a direct negative bearing on the resilience of the ex-FAR, Interahamwe and Rwandan civilian refugees in the DRC to return to Rwanda. If the RPF had adopted a reconciliatory approach in its leadership from the initial stages, a number of the Hutu refugees, including *génocidaires* would perhaps have embraced the spirit of reconciliation and returned to Rwanda, thereby diffusing the tension in eastern DRC.

The uncertainty brought about by the failure of the RPF to establish strategic victory manifested itself further in August 1995 when the Prime Minister, Twagiramungu and the Interior Minister, Seth Sendashonga, a Hutu but RPF follower resigned due to fear of retribution, and immediately fled into exile. In the same month the Minister for Justice, Alphonse Nkubito resigned and died two years later (Reyntjens, 2004: 180). Sendashonga was later assassinated in Nairobi. Several government ministers, high-ranking civil servants, judges, army officers, journalists and civil society activists left Rwanda in a continuous stream, as described by Reyntjens (2004: 182–187). These internal occurrences seemed to worsen the tension that the ex-FAR/Interahamwe was exerting on the RPF. In retaliation, the RPF evolved their strategy to encompass the overthrow of Mobutu not only for having supported Habyarimana, but also for the support that Mobutu had extended to the ex-FAR/Interahamwe.

The abrogation of the Arusha Accords[2] ended with a tactical victory by RPF, which, however, fell considerably short of transforming its success into sustainable peace. As the RPF regime grappled with this challenge, the Interahamwe and the ex-FAR inhabiting the refugee camps re-organised themselves to mount a revenge by recapturing power in Kigali, with the support of Mobutu. Whitaker writes that as a close ally of former Rwandan president Juvenal Habyarimana, Mobutu, whose regime had been weakened considerably due to his loss of Cold War Western allies and increased internal pressure from opposition parties clamouring for democratisation, took advantage of the failure by Rwanda to acknowledge the need to negotiate and reconcile with the former regime, including the return of refugees. He sought the support of Rwandan refugees, the ex-FAR and Interahamwe to fight against the new RPF regime in Kigali. He also offered shelter and protection to officials of the former Rwandan government. Mobutu's government allowed the Interahamwe and ex-FAR to train near the refugee camps for an eventual return by force to their home country, while at the same time facilitating the acquisition of arms for them as well as permitting the use of Goma airport for delivery of weapons destined to the remnants of FAR that were still in Rwanda (Saint-Exupery, 1998: 4).

To put their threats into practice, the ex-FAR/Interahamwe adopted an insurgency strategy that entailed recruiting and re-arming, such that by 1995 the group established what it called "Government of Rwanda" in exile (Koen, 2002: 11). From the refugee camps the extremists launched small-scale insurgency into Rwanda with the tacit support of Mobutu's regime. Orth (2001: 76) depicts the ex-FAR insurgency as a case of reversed roles: "This action signified a juxtaposition of roles: the counterinsurgent Hutu-dominated government and its military, the FAR, becoming insurgents; and the guerrilla RPF leading a broad-based government of national unity and its military, RPA, becoming the counterinsurgents." Noticing this, the RPF intensified its counter-insurgency operations that they had started immediately after the genocide stopped.

The RPF covertly infiltrated refugee camps in Zaire, in which it identified and killed those it targeted. In retrospect, the ex-FAR and Interahamwe intensified insurgency activities in Rwanda. They infiltrated the Western *Préfectures* (provinces) of Cyangugu, Kibuye and Gisenyi from early 1995 (Reyntjens, 2011: 133) and terrorised villagers. This worsened the situation beyond the expectations of the RPF, prompting the then-vice-president Paul Kagame to openly make known his strategic war plan. While on a visit to the United States in August 1996, Kagame informed Americans that he was on the verge of acting against ex-FAR attacks being launched from Zaire. According to Kagame, his intelligence network had indicated that ex-FAR were planning a large-scale offensive against Rwanda from Goma and Bukavu, and the only way was to disrupt this by an immediate invasion of Zaire.

It was after Kagame's declaration of his intentions that the RPF devised a more robust counter-insurgency strategy that would later lead to invasion of the refugee camps in Zaire and the eventual overthrow of Mobutu. In this regard, the RPF initial political objective of attaining total victory against the ex-FAR/Interahamwe evolved into the objective of deposing Mobutu. In order to achieve this goal, the RPF had to ensure two things. Firstly, it needed to conceal its image as an aggressor by devising a plausible reason to justify its entry into Zaire. Secondly, the RPF sought to ensure that its plans to attack the refugee camps were a top secret so as not to alert the *génocidaires* who would otherwise escape. To achieve the latter goal, the RPF devised a strategy based on the classical military actions of deception and surprise. The opportune moment for the implementation of this plan came when the Governor of South Kivu called upon all the Tutsi to leave the country or face retaliation at the height of the conflict in 1996. Fearing the consequences of the warning, the Banyamulenge coalesced around the AFDL-appointed leader Laurent-Désiré Kabila, who, with the Rwandan and Ugandan support, initiated a rebellion against the Mobutu regime (Rogier, 2003: 5). As stated earlier, the loss of political and military support to Mobutu from his Western allies had further rendered him vulnerable to his erstwhile enemies (Baregu, 2006: 60).

With the support of Rwanda and Uganda, Kabila marshalled the military strength under his political party, PRP to challenge Mobutu. Prunier (2009: 114–117) clarifies the regional aspects of the conflict when he writes that after creating PRP, Kabila thought of continuing a classical guerrilla struggle against the Mobutu regime, but had a limited number of weapons, and his fighters came largely from one tribe: the Bambembe. Rwanda sought the backing of Banyamulenge, in a bid to reinforce Kabila. When Rwanda sought for ways of neutralising its enemies inhabiting refugee camps across in the

DRC they found Laurent-Désiré Kabila, a Congolese from Katanga, and a former Lumumbist (a Katanga separatist group of the 1960s), who had for many years tried to topple Mobutu in vain (Schatzberg, 1997: 75). Museveni is said to have settled on Kabila as the best choice for the task. The strategy was that Kabila would champion the citizenship course for the Banyamulenge, so as to give the issue a "national" outlook. Once Kabila had been co-opted in the plot Kagame and Museveni proceeded to the next strategy: the establishment of the AFDL – the rebel group that Kabila would lead. The stage had been set for Rwanda's invasion of Zaire, and hence the extension of its civil war into Zaire.

The changes in the strategic global environment

This contributed a great deal to the weakening of the political regime in Zaire. The main changes were the ending of the Cold War and the democratic demands that accompanied it. Although this was largely a proximate (external) cause, its impact was much more intensive on the internal structures of Zaire which were already faced with similar (democratisation) demands owing to the lengthy dictatorial regime of Mobutu. It is therefore instructive to trace the major events that marked Congo's political regime during the Cold War, as the basis for understanding how the end of the Cold War impacted on the country's regime (Arnold, 2008: 767–771). The civil war and the successive coup of 24 November 1965 that brought Mobutu to power, ousting Kasavubu and Tshombe, was backed by America's Central Intelligence Agency (CIA) to ensure a Congo leader who would safeguard Western interests (Nzongola-Ntalaja, 2002: 20–23). The Mobutu regime was therefore entirely a product of the Cold War. Arnold states that as an externally backed autocracy, the Mobutu regime was a pure product of the Cold War, based on the strategic calculation of Western powers that leaders with no social or political base were preferable to those with strong national constituencies, to which they were accountable (Arnold, 2008: 770–771).

Mobutu did not therefore strive to popularise and consolidate his leadership across Congo because after all he had the backing of the West. This was one of Mobutu's biggest miscalculations that later led to his gradual alienation from the rest of the country and the fall of his regime in the early 1990s when the Cold War ended. Hochschild (1998) and Baregu (2006) trace the beginning of Zaire's troubles to this period. They contend that in addition to ethnicity and conflicts that were taking place at the same time in Rwanda and Burundi, the interplay among the seemingly endless supply of mineral resources, the greed of multinational corporations desperate to cash in on the DRC's wealth, raged on.

Whitaker (2003: 10) stresses that Mobutu's downfall, which also increased the vulnerability of Zaire's state security, was as a result of the changes in both political and economic environments that started in the 1980s, and which was marked by the regime's unaccountable misappropriation of national resources. Mobutu's actions eroded the country's economy, causing churches and civil society groups to increase their pressure for political reform. As was the situation elsewhere in Africa, the West no longer needed Mobutu's allegiance and therefore they cut off the support that he had enjoyed during the Cold War era. According to Prunier (1995: 376), Mobutu was "harassed for human rights violations, denounced by the media for his corruption, marginalised by the International Monetary Fund (IMF) for non-repayment of loans and even at one point banned from Europe when his old ally France refused him an entry visa". In this context, the situation in Zaire blended with the ending of the Cold War in the early 1990s, the period when the Rwandan civil war was erupting (see proximate causes of the war). The early 1990s were also marked by a wave of democratisation and multi-party politics. This fused with the existing challenges to Mobutu, who instead of building a national consensus around the process, attempted to manipulate the process through dishonesty, thereby aggravating the situation.

Conclusion

The central argument of this chapter is that the spread of war in the DRC was as a result of a combination of core and proximate causes. The chapter illustrated how the external (spatial) proximate events (that took place in countries that are neighbours of the DRC) merged with the internal (core) factors of instability already present in the DRC, to trigger the civil war that started in 1996. It also presented a conceptual discussion of the core and proximate causes of the war in the context of their spatial spread. The emphasis here is that although in the case of the DRC civil war some of the proximate causes originated from the country's immediate neighbours, this may not always be the case in other conflicts, as it is also possible that actions by actors from beyond the neighbourhood are equally important. Because the war eventually spread across the Great Lakes Region, the chapter examined first what constitutes this region in terms of the common characteristics that facilitate the ease with which conflict occurs. It then examined the core causes of the conflict, followed by the proximate causes. It concludes, shortly, by presenting a case of how the proximate causes played on the weak political fibre established by the core causes, to trigger the civil war that started in 1996.

The first core cause was the change in strategic environment. The chapter has argued that President Mobutu's overreliance on Western powers during the Cold War era was in itself a weakness, whose effects became apparent with the end of the Cold War. This left Mobutu's regime too weak and vulnerable to both internal political opposition forces in the DRC, and to invasion by the RPA, who found it easy to topple Mobutu. The second core cause was the global democratisation demands orchestrated by the advent of multi-party politics in the early 1990s. This wave of political change found Mobutu's regime already under pressure from the changed strategic environment. In addition to this, opposition parties sprung up within the DRC, thereby exerting more pressure on Mobutu's corrupt and ethnically divided leadership.

The third cause was the ethnic divisions amongst the Congolese in the east of the country, primary among them being the politicisation of the issue of the Rwandaphone citizenship. Rwanda exploited the cross-border ethnic ties amongst the Kinyarwanda speaking people in Rwanda, Burundi and the DRC to mobilise resistance against Mobutu in the region, while at the same time exploiting the citizenship issue of the Banyamulenge to mobilise a rebellion against Mobutu, in revenge for the latter's support of President Habyarimana. The endemic hostility between Hutu and Tutsi groups spread across Rwanda, Burundi and eastern DRC at historic levels. The ease with which the conflict spread across the borders of the three countries is owing to cross-border ethno-political ties binding the inhabitants of the three countries. It was within this context that Rwanda and its allies invaded the DRC and sparked a long-drawn war whose conduct and implications continue to be felt today.

The Great Lakes Region has a diverse combination of cultural and ethnic trends. Ethnicity has the potential to be either a source of strife and violence in heterogeneous communities or a cause of unity in diversity. In the latter case, it can be used to promote national and regional cohesion and therefore form a solid foundation for economic and political strength. Sadly, however, ethnic relations can lead to competition for political space, territory, opportunities and inter-ethnic identity competition. This leads to prolonged hatred and conflict and bigotry, that, as in the case of the GLR, can escalate into civil wars. Whereas ethnicity has been a major core and proximate cause of cross-border conflict in GLR as noted in this chapter, there is great potential to transform this negative historical trend into a solid inter-country/intra-regional economic strength hinged on these linkages. For this to happen, it is recommended that firstly, the political elite in the GLR capitalise on the cross-border linkages to encourage unhindered interaction between citizens of neighbouring countries as a way of fostering collegiality from the grassroots where the ethnic differences are mostly manifested at first. Secondly, the regional actors need to urgently address the ever-present question of the nationalities of the ethnicities that remain under contestation in the region.

This should be handled in the realm of the advantages these people bring to wherever they are, and not be used as a "tool" to sharpen ethnic differences. This can best be approached by the region adopting policies on free movement and co-existence of people based on a common regional identity and not on conflicting ethnicity. Thirdly, the long-term reconciliation and development in the region will be further enhanced if objective transitional justice mechanisms to deal with historical injustices that are at the centre of the core and proximate causes of the cross-border cleavages in these countries can be established. Such mechanisms should also prioritise issues of gender violence and related atrocities meted against women and children either directly through sexual abuse or as a collateral to armed violence. These issues have not been covered in this chapter given its limited scope but warrant further research and discussion.

Notes

1 See, "Key Concepts in Conflict and Peace", at: https://extranet.creativeworldwide. com/CAIIStaff/Dashboard_GIROAdminCAIIStaff/Dashboard_CAIIAdminDataba se/resources/ghai/understanding.htm (accessed on 27 April 2018).
2 Both Rwanda Patriotic Front (RPF) and the Movement Révolutionnaire National pour le Développement (MRND) violated the 1993 Arusha Accords that they had both signed. Both sides re-armed discretely and engaged in subversive activities and killings against each other. Eventually President Habyarimana was assassinated and the civil war that ensued between the Hutu and Tutsi led to the 1994 genocide (Jones, 1999: 141).

References

Agnew, J. 2008. Borders on the Mind: Re-framing Border Thinking. *Ethics & Global Politics* 1(4): 175–191.

Arnold, M. 2008. Intervention. In C. A. Snyder (Ed.), *Contemporary Security and Strategy*. New York: Palgrave Macmillan.

Banégas, R. 2008. Rethinking the Great Lakes Crisis: War, Violence and Political Recompositions in Africa. In J-P. Chrétien and R. Banégas (Eds), *The Recurring Great Lakes Crisis. Identity, Violence and Power*. London: Hurst & Co., pp. 1–25.

Baregu, M. 2006. Congo in the Great Lakes Conflict. In M. Gilbert Khadiagala (Ed.), *Security Dynamics in Africa's Great Lakes Region*. Boulder, CO: Lynne Rienner Publishers.

Bonner, R. 1994. French Establish a Base in Rwanda to Block Rebels. *The New York Times*, 5 July.

Burton, J. 1990. *Conflict: Resolution and Prevention*. New York: St. Martin's Press.

Chajmowicz, M. 1996. Kivu: les Banyamulenge enfin à l'honneur. *Politique Africaine* 64 (December): 115–120.

Chrétien, J-P. and Banégas, R. 2008. *The Recurring Great Lakes Crisis. Identity, Violence and Power*. London: Hurst & Co.

Clark, J. 1995. Ethno Regionalism in Zaire: Roots, Manifestations and Meaning. *Journal of African Policy Studies* 1(2): 33–34.

Clark, J. 1998. The Nature and Evolution of the State in Zaire. *Studies in Comparative International Development* 32(4): 3–23.

Clark, J. (Ed.) 2007. *The African States of the Congo War*. Kampala: Fountain Publishers.

Coleman, P. 2011. Intractable Conflict. In M. Deutsch and P. Coleman (Eds), *The Handbook of Conflict Resolution: Theory and Practice*. San Francisco: Jossey-Bass.

Collins, R. 2008. *Violence: Micro-Sociological Theory*. Princeton, NJ: Princeton University Press.

Daley, P. 2006. Challenges to Peace: Conflict Resolution in the Great Lakes Region of Africa. *Third World Quarterly* 27(2): 303–319.

Declaration of the Summit of Heads of State and Government of the Member States of ICGLR on the Security Situation in Eastern Congo. 2013. September. www.sta

tehouse.go.ug/media/news/2013/09/06/declaration-summit-heads-state-and-governm ent-member-states-icglr-security-sit.

Edmonds, M. , Mills, G. and McNamee, T. 2009. Disarmament, Demobilization, and Reintegration and Local Ownership in the Great Lakes: The Experience of Rwanda, Burundi, and the Democratic Republic of Congo. *African Security* 2(1): 29–58.

Evans, G. 1997. *Responding to Crises in the African Great Lakes*. New York: Oxford University Press.

Ewald, J. 2004. *A Strategic Conflict Analysis for the Great Lakes Region*. Gothenburg, Sweden: Gothenburg University.

Galtung, J. 1969. Violence, Peace and Peace Research. *Journal of Peace Research* 6(3): 167–191.

Harff, B. and Gurr, T. 2004. *Ethnic Conflict in World Politics*. Boulder, CO: Westview Press.

Harvey, D. 2004. *The New Imperialism*. Oxford: Oxford University Press.

Hochschild, A. 1998. *King Leopold's Ghost: A Story of Greed, Terror and Heroism in Colonial Africa*. Basingstoke: Macmillan Publishers.

Huntington, S. 1993. *The Third Wave. Democratisation in the Late Twentieth Century*. Oklahoma: University of Oklahoma Press.

Jones, B. 1999. The Arusha Peace Process. In H. Adelman and A. Suhrke (Eds), *The Path of a Genocide. The Rwanda Crisis from Uganda to Zaire*. New Brunswick, NJ: Transaction, pp. 131–156.

Kabanda, M. 2008. Rwanda, the Catholic Church and the Crisis. Autopsy of a Legacy. In J-P. Chrétien (Ed.), *The Recurring Great Lakes Crisis, Identity, Violence and Power*. London: Hurst & Co.

Kalron, N. 2010. The Great Lakes of Confusion. *African Security Review* 19(2): 25–37.

Keane, F. 1995. *Season of Blood*. New York: Penguin Books.

Koen, V. 2000. Identity and Insecurity: The Building of Ethnic Agendas in South Kivu. In D. Ruddy and G. Jan (Ed.), *Politics of Identity and Economics of Conflict in the Great Lakes Region*. Brussels: VUB Press.

Koen, V. 2002. Citizenship, Identity Formation & Conflict in South Kivu: The Case of the Banyamulenge. *Review of African Political Economy* 29(93–94): 499–516.

Lake, D. and Rothchild, D. 1996. Containing Fear: The Origins and Management of Ethnic Conflict. *International Security* 21(2) (Autumn): 41–75.

Lemarchand, R. 1996. *Burundi: Ethnic Conflict and Genocide*. Cambridge: Cambridge University Press.

Marysse, S.De Herdt, T. and Ndayambaje, E. 1994. *Rwanda. Appauvrissement et ajustement structurel*. Cahiers Africains 12. Paris: L'Harmattan.

Maslow, A. H. 1973. *The Farther Reaches of Human Nature*. Harmondsworth, UK: Penguin Books.

McGreal, C. 1996. Muganga Uncovers the Crushed Militias' Blueprints for Battle. *The Guardian*, 18 November.

Moller, B. 2003. *Conflict Theory*. Working Paper No 122. Research Center on Development and International Relations (DIR). Denmark: Aalborg University.

Mutta, P. 2004. New Horizons: The Great Lakes Region of Africa Political, Social and Military Stabilisation. Paper presented at the Canadian Forces College, 30 April.

Nzongola-Ntalaja, G. (Ed.). 2002. *The Congo: From Leopold to Kabila. A People's History*. New York: Zed Books.

Orth, R. 2001. Rwanda's Hutu Extremist Genocidal Insurgency: An Eyewitness Perspective. *Small Wars & Insurgencies* 12(1): 76–109.

Pottier, J. and Fairhead, J. 1991. Post-Famine Recovery in Highland Bwisha Zaire. *Africa* 61(4): 437–470.

Pourtier, R. 1996. La guerre au Kivu: Un conflict multidimensionel. *Afrique Contemporaine* 180(10/12): 15–13.

Prunier, G. 1995. *The Rwanda Crisis 1959–1994. History of a Genocide*. London: Hurst and Co.

Prunier, G. 2009. *From Genocide to Continental War, The "Congolese" Conflict and the Crisis of Contemporary Africa*. London: Hurst Publishers.

Reyntjens, F. 1996. Constitution-making in Situations of Extreme Crisis: The Case of Rwanda and Burundi. *Journal of African Law* 40(2): 234–242.

Reyntjens, F. 2004. Rwanda, Ten Years On: From Genocide to Dictatorship. *African Affairs* 103: 177–210.

Reyntjens, F. 2006. Governance and Security in Rwanda. In M. Gilbert Khadiagala (Ed.), *Security Dynamics in Africa's Great Lakes Region*. Boulder, CO: Lynne Rienner Publishers.

Reyntjens, F. 2011. Waging (Civil) War Abroad. In S. Straus and L. Waldorf (Eds), *Remaking Rwanda, State Building and Human Rights after Mass Violence*. Madison: University of Wisconsin Press.

Rickard, N. 1999. The Causes of War. *The Journal of Conflict Studies* 19(1): 5–43.

Rogier, E. 2003. *Cluttered with Predators, Godfathers and Facilitators: The Labyrinth to Peace in the Democratic Republic of Congo*. The Hague: Netherlands Institute of International Relations.

Rufin, J-C. (1996). Les économies de guerre dans les conflits internes. In F. Jean and J-C. Rufin, *Economie des guerres civiles*. Paris: Hachette.

Ruhimbika, M. 2001. *Les Banyamulenge (Congo-Zaïre) entre deux guerres*. Paris: l'Harmattan.

Saint-Exupéry, P.De 1998. La France aurait fourni des armes au Rwanda pendant le genocide. *Le Monde*, 13 January.

Sambanis, N. 2003. *Using Case Studies to Expand the Theory of Civil War*. CPR Working Papers, Paper No. 5.

Schatzberg, M. 1997. *The Dialectics of Oppression in Zaire*. Indianapolis: Indiana University Press.

Tegera, A. 1995. La réconciliation communautaire: le cas des massacres au Nord-Kivu. In A. Guichaoua (Ed), *Les crises politiques au Burundi et au Rwanda (1993-1994). Analyses, faits et documents*. Villeneuve d'Ascq, Université des Sciences et Technologies de Lille.

United Nations 2000. *We the Peoples*. United Nations Millennium Report. New York: United Nations Department of Public Information.

Uvin, P. 1998. *Aiding Violence. The Development Enterprise in Rwanda*. West Hartford, CT: Kumarian Press.

Van Houtum, H. 2012. The Geopolitics of Borders and Boundaries. *Geopolitics* 10(4): 672–679.

Van Hoyweghen, S. and Koen, V. 2000. Ethnic Ideology and Conflict. In D. Ruddy and G. Jan (Eds), *Sub-Saharan Africa: The Culture Clash Revisited in Politics of Identity and Economics of Conflict in the Great Lakes Region*. Brussels: VUB Press.

Walter, S. 1998. Waiting for "the Big One": Confronting Complex Humanitarian Emergencies and State Collapse in Central Africa. *Small Wars & Insurgencies* 9(1): 72–101.

Whitaker, B. 2003. Refugees and the Spread of Conflict: Contrasting Cases in Central Africa. *Journal of Asian and African Studies* 38(2): 211–231.

Willame, J-C. 1995. *Aux sources de l'hecatombe rwandaise.* Cahiers Africains 14. Paris: L'Harmattan.

Willame, J-C. 1997. *Banyarwanda et Banyamulenge: Violences Ethniques et Gestion de l'Identitaire au Kivu.* Brussels: Institut Africain – CEDAF.

6 The cross-border dimension of intrastate conflicts in Africa

An analysis of the Great Lakes region and Mano River

Jackson A. Aluede

Introduction

Since the end of the Cold War there has been unprecedented increase in the number of violent conflicts in different parts of the world. Interestingly, the majority of the conflicts have been intrastate conflicts or civil wars – a remarkable departure from the ideological and proxy wars that characterised conflicts during the Cold War era. In Africa, following the collapse of the Berlin Wall in 1989, previously frozen conflicts in the continent erupted in the form of intrastate conflicts in countries such as Liberia, Sierra Leone, Chad, Rwanda, Burundi, Sudan and the Democratic Republic of Congo (DRC) (Cilliers and Schünemann, 2013). Similarly, as in other parts of the world, the collapse of communism fuelled the resurgence and the politicisation of ethno-nationalism, tyranny, marginalisation, among others factors which turned out to be the most important parameters of intrastate conflicts in Africa (Olayode, 2016: 242).

The effects of intrastate conflicts in Africa are alarming. Ramsbotham, Woodhouse and Miall (2005: 72) found out how citizens of Angola, Eritrea, Liberia, Mozambique, Rwanda, Somalia and Sudan, were forced to flee their countries following the outbreak of intrastate conflicts in the 1990s. In West Africa, between 1989 and 2000, an estimated 800,000 fatalities were recorded in the civil wars in Liberia and Sierra Leone (Marc, Verjee and Mogaka, 2015: 4). In parts of Eastern and Central Africa, intrastate conflicts in Rwanda, Burundi and DRC have resulted in approximately 7 per cent of the refugee population, causing massive dislocations and casualties among civilians (Lemarchand, 1997).

Intrastate conflicts in the post-Cold War period occurred more in developing and heterogeneous societies, divided along ethnic, religious and ideological lines. This was evidenced in Eurasia, former Soviet Union and Yugoslavia territories in which the breakup of states occurred along lines of situational ethnicity, with religion, language or regional or clan affiliation serving in most cases as the markers of identity (Hopmann, 1999). In Africa, opinions are divided among scholars on the factors or causes of intrastate conflicts. Some like Mahmood Mamdani (2002, 2005) blamed intrastate conflicts in Africa, particularly in the Great Lakes region, on the failure of the colonial and post-colonial governments in the region to manage the citizenship and identity question among some ethnic groups, particularly the Hutus and Tutsi in the region. Adebayo Adedeji (1999), distances himself from the ethnic perspective as the major cause of intrastate conflicts in the continent. He argues that the problem with this type of approach or analysis is the lack of clarity and con-sciousness as to the exact make-up of the ethnic or tribal phenomenon (1999: 9). On their part, Collier and Hoeffler (2004) identify greed and grievance as some of the causes of the outbreak of civil wars or intrastate conflicts in Africa as well as other parts of the world, particularly, the role of rebel groups. They argue that:

> Greed and misperceived grievance have important similarities as accounts of rebellion. They provide a common explanation – "opportunity" and "viability" describe the common conditions sufficient for profit-seeking, or not-for-profit, rebel organisations to exist.
>
> (Collier and Hoeffler, 2004: 565)

The outbreak of intrastate conflicts in Africa since the end of the Cold War has equally been attributed to the failure of some states in the continent to perform some fundamental tasks for their citizens. Jean-Paul Azam (2001: 442) argues that the occurrence of civil conflict in Africa is intimately related to the failure of governments to deliver the type of public expenditure that the people want, that is, public expenditure with a strong redistributive component such as health and education. Other factors that have con-tributed to the eruption of intrastate conflicts in Africa in general, and the Great Lakes and Mano River regions in particular, include the colonial legacy, uneven distribution of land and mineral resources, political victimi-sation, ethnic domination, the absence of internal democracy and good governance, rule of law, equity and religious intolerance (Shyaka, 2008; Sawyer, 2004; Lemarchand, 1998). Similarly, the destructive impacts of intrastate conflicts in Africa particularly, with respect to forced migration, pillage of mineral resources, disruption of the democratic process, infra-structural development and economic stability have received enormous attention in the literature (Akokpari, 1999; Tom, 2014; Onuoha, 2004; Aning and Pokoo, 2017; Rwengabo, 2014). Nevertheless, there exists a gap in cross-border dimensions of intrastate conflicts in Africa, particularly in the Great Lakes and Mano River regions including Burundi, Rwanda and the DRC, and Liberia and Sierra Leone, respectively.

This chapter focuses on the cross-border dimension of intrastate conflicts. It analyses the state and nature of Africa's borders in general, and those of the Great Lakes and Mano River regions in particular, to ascertain the extent to which the borders have impacted on intrastate conflicts across the borders of both regions. Since the 1990s, intrastate conflicts in the Great Lakes and Mano River regions have been characterised by strong ethnic attachment and contestations for control of mineral resources to support the conflicts – aided by the fluidity of the borders, thanks to their poor demarcations and weak border control of countries in both regions. The chapter also examines efforts to address the border problematic facing both regions.

Conflict in the post-Cold War era

Charles Hauss (2010) reveals that both interstate and intrastate conflicts have dominated the nature and patterns of conflicts since 1945, and that less attention had been given to intrastate conflict until end of the Cold War. In his words:

> Since the end of the Cold War, our attention has been riveted on civil and other primarily intrastate wars. Interstate wars have become the exception rather than the rule, especially those between major powers that have at the heart of traditional international relations theory and analysis. Depending on how you count, at least 80 per cent of the violent conflicts since 1945 have been intrastate. We just did not focus on them as such until the 1990s because we tended to see them as proxy battles in the Cold War, rather than disputes to be analysed in their own right.
>
> (Hauss, 2010: 11)

Interstate conflicts occurred between states following irreconcilable differences that impinge on their national interest. Some of the reconcilable differences are border dispute, dispute over mineral resources, territorial invasion and the fight against terrorism (Zeleza 2008; Cameron, 2005). The outbreak of interstate conflicts in different parts of the globe has resulted in thousands of deaths, ruined the economy of the belligerents and contributed to forced migration, despite lesser occurrences since the 1990s. Intrastate conflicts on the other hand are protracted, have complex causes, and, despite the use of low-tech weaponry, have led to millions of deaths and forced migration and internally displaced persons.

The Cold War period witnessed an increase in proxy wars in different parts of the world in which warring factions were clandestinely supported by either the USA or the Soviet Union, with the aim of promoting their ideological interest (McWilliam and Piotrowski, 1997). Following the collapse of the Soviet Union in 1989, it was expected that armed conflicts would reduce significantly, as one of the major actors fuelling proxy wars had disintegrated. Unfortunately, conflicts around the world did not end. Instead, intrastate armed conflicts emerged in different parts of the world. Wallensteen and Sollenberg (2001: 65) point out that from 1989 to 2000 the number of conflicts around the world was 111. Of this figure, 104 were intrastate. Specific instances of interstate conflicts during the period included the US-led coalition against Iraq in 1990 and the US invasion of Iraq and Afghanistan in 2003, the conflicts in India and Pakistan of 2001–2003, and in Djibouti and Eritrea in 2008 (Themner and Wallensteen, 2012).

Since the end of the Cold War, intrastate conflicts have increased significantly in different parts of Africa – resulting in African and Western intervention to resolve the conflicts. According to Alfred Nhema:

> Since the demise of the Soviet Union, the nature of warfare in Africa has shifted to mostly intrastate conflicts … . Concomitantly with this change has been a lowering of the geographical strategic importance given to Africa by the Western powers. Hence, in the presence of conflict and failing states, the West has either failed to respond or responded with some reluctance, as exemplified in the situations prevailing in such countries as Somalia, Rwanda, Burundi and the Democratic Republic of Congo (DRC).
>
> (Nhema, 2008: 3)

The Mano River and the Great Lakes regions feature prominently among areas experiencing intrastate conflicts in the last decades of the twentieth century. The trend continued into the first decade of the twenty-first century. And according to the Uppsala Conflict Data Program (UCDP) authored by Themner and Wallensteen (2012), from 2001 to 2010, the number of armed conflicts across the world was 69, dominated by intrastate conflicts. Of the 69 armed conflicts active in 2001–2010, 27 (or 39 per cent) were fought in Africa. Intrastate conflicts during this period had subsided in Mano River region, but not so in the Great Lakes region, particularly in the DRC.

The rest of the chapter is divided into four sections. The first section discusses concepts and patterns of conflicts in the post-Cold War era and the nature of Africa's borders. The second section gives an overview of the cross-border dimension of conflicts in the Great Lakes and Mano River regions. The third section engages in an examination of intrastate conflicts and the cross-border dimensions through case studies of the two regions. The fourth section discusses some of the ways in which attempts are made to resolve the intrastate conflicts that have cross-border implications in each of these regions, and the last section concludes with recommendations based on the discussion in the previous sections.

Borders and intrastate conflict in Africa: some theoretical considerations

According to Stephen Walt (2005: 23), theory is an essential tool for explaining the causes, trends and impact of different issues and policies. To this effect, some scholars in the humanities and social sciences in their attempt to understand the dynamic of conflicts in the international system as well as unravel the factors responsible and in order to find a lasting solution to intrastate conflicts in particular have employed theories to justify their submissions or arguments. One such scholar is Edward Azar (1990). His theory of protracted social conflict (PSC) provides insight into the root of conflicts during the Cold War period and how to resolve them. Azar argues that critical factors responsible for PSC in some countries such as Lebanon, Sri Lanka, the Philippines, Nigeria, South Africa, Northern Ireland and some other countries, was the prolonged and often violent struggle by communal groups for such basic needs as security, recognition and acceptance, fair access to political institutions and economic participation (Ramsbotham, Woodhouse and Miall, 2005). Furthermore, he argued that PSC emphasised that the sources of conflicts lay predominantly within (and across) rather than between states, with four clusters of variables identified as preconditions for transformation to high levels of intensity. The four clusters are communal content, deprivation of human needs, state endowed with authority to govern and use force, and international linkages. Azar concludes that the role of the state with respect to PSC is to satisfy or frustrate basic communal needs, thus preventing or promoting conflict (Azar, 1990). Azar's theory of PSC applies to the conflicts in the Great Lakes and Manor River regions.

In addition, this chapter follows Gurr's perspective on ethnicity. According to Gurr (2009) ethnicity has to do partly with primordial interpretation of issues which recognises the differences among humankind and societies based on their race, language, customs, norms and civilisation. These differences have been manipulated by political leaders for political and material benefit (Gurr, 2009). Ethnicity has been responsible for several conflicts in different parts of the world. For instance, in former Soviet Union and Yugoslavia territories, it was responsible for the clashes between the armies of Croatia, Serbia and Slovenia, and the battle that pitted Bosnia's Croats, Muslims and Serbs against each other (Sadowski, 1998: 12). In Africa, ethnicity has contributed to the outbreak of intrastate conflicts in different parts of the continent, including the Great Lakes and Mano River regions. For example, it has influenced conflicts between the Hutu and Tutsi, in Rwanda, Burundi and the DRC (Ewald, 2004).

Further, the notion of transnationalism is also utilised in this chapter. Transnationalism refers to cross-border processes involving diverse individuals and groups or multiple ties and interactions linking people and institutions across the borders of nation-states (UNESCO, 2013). Transnationalism provides an explanation for the phenomena that transpire across the borders of nation-states – migration, social, economic and political processes including social movements, governance and politics, terrorism, political violence, and organised crime among others (Faist, 2009: 11). It covers the activities of transnational communities whose people engage in mutual interaction across national boundaries and are orientated around a common project or "imagined identity" (Djelic and Quack, 2010). In the Great Lakes and Mano River regions, the activities of rebel leaders and warlords in pillaging and plundering the mineral resources of neighbouring territories are driven by transnational interest, aimed at maintaining their hold over territories under their control. These individuals and groups operating on both sides of the borders with their associates, take advantage of the weak and porous state of the borders to achieve transnational goals and for economic gains.

Transnationalism also explains forced migration among civilians fleeing intrastate conflict in the Great Lakes and Manor River regions.

Colonial legacy, porous borders and cross-border implications of intrastate conflicts in Africa

Borders all over the world have played a very significant role in influencing the socio-cultural, economic, political and security wellbeing of people residing in borderlands and the larger society of nation-states. This fact has been echoed in several works by scholars in the field of border studies (Zartman, 2010; Wilson and Donnan, 2012; Anene, 1970; Asiwaju 1984). The porous and artificial nature of the borders of nation-states and their border policies, particularly in the era of globalisation have likewise raised some fundamental questions regarding the appropriate border policy that would enable nation-states to benefit from globalisation and reduce, if not eliminate, its security implications across their borders (Rajaee, 2000: 25; van Schendel and Abraham, 2005). The development elicited the open and closed border discourse or debate among scholars who argued for and against its economic and security implications and other border problematics such as cross-border crime, illegal smuggling of arms and ammunition, mineral resources and trafficking of persons across the border.

Africa like other parts of the world has had her share of border problematics following the end of colonialism. The continent's border problematic manifested owing to its porous nature, poor border management, location of ethnic groups and mineral resources across the borders between nations, corruption and ineffectiveness of border personnel and inarticulate border policy of governments (Asiwaju, 2003). However, the bane of African border challenges is traced to the infamous Berlin Conference of 1884/85, where the respective European imperialist powers partitioned the continent among themselves without the consent of the people, or consideration of the repercussions of their actions on the continent and its peoples (Hargreaves, 1984).

The partitioning of the African continent sowed the seeds of border-related challenges that have continued to confront post-independence African states. Ieuan Griffiths, commenting on the effect of the Berlin conference of 1884/85, concludes that: "The inherited political geography of Africa is as great an impediment to independent development as her colonially based economies and political structures" (Griffiths, 1986: 204).

The new boundaries created by imperialist European powers removed the continent's ancient landmarks, and separated kin and kith in different parts of Africa. The adverse effect of the arbitrary boundary demarcation continued into post-independence Africa and equally led to the emergence of ethnic or national minorities, with the resulting questions of irredentism and the alarming practice of ethnic cleansing, border disputes over mineral resources, smuggling and cross-border criminal activities. Some of the effects the colonially inherited borders have had on the continent were exacerbated by the nature of Africa's borders in the post-colonial era. This is the submission of Francis Nguendi Ikome, who argues that:

> Africa's borders are very porous because of a lack of proper demarcation and delimitation. This has been identified as the principal reason for the ease with which governance-related national conflicts in individual states have spilled over to entire regions, as has been the case in the Great Lakes region, West Africa and the Horn of Africa. Significantly, many intra-state conflicts in Africa have been sparked by the forceful fusion of incompatible national groups into one state by the imposition of artificial boundaries by colonial powers.
>
> (Ikome, 2012: 1)

Aside from the above fact, there are other challenges across Africa linked to the partitioning which the leaders on the continent have continued to encounter after the end of colonialism. For instance, Togo and Ghana, shortly after independence, were involved in the irredentist squabble over the Ewe people along their border. In the Great Lakes region, the Burundi and Rwanda genocide of the 1990s and the pockets of cross-border ethnic attacks between the Hutus and Tutsis were partly the manifestation of the arbitrary demarcation of the Berlin arrangement (Zartman, 1965; Tom, 2014).

Apart from the colonial legacy, Africa's post-colonial leaders on their parts have equally contributed to border challenges. Several of them fail to articulate comprehensive and effective border policies to address the many border challenges facing them. In addition, other complicating factors include the absence of enough border personnel to police the border, and an inability to clearly demarcate their borders – pursuant to hostile border relations with their immediate neighbours due to fear and suspicion and protracted legal cases over boundary demarcation between African countries (Asiwaju, 2003). The most worrisome African border problematic that has continued to pose serious threats to the continent is the porous nature of her borders and the separation of ethnic groups across nations sharing boundaries.

The porous nature of the continent's borders has been exploited by groups and individuals particularly those involved in cross-border crime, smuggling and other forms of illicit enterprise. It is equally interesting to note that the porous nature of Africa's borders has given impetus to the cross-border dimension of intrastate conflicts in the continent. The porous nature of Africa's borders has enabled the spillover of intrastate conflicts across borders along ethnic lines (Ikome, 2012). This was the case in the Great Lakes region during the conflict between the Hutus and Tutsis in Rwanda and Burundi. Rebel groups have equally exploited the porous borders of the continent to cause intrastate conflicts to spill over into another territory, with the motive to allow the rebels to exploit mineral resources in order to prolong the conflict and to enrich themselves. The experiences of Liberia and Sierra Leone attest to this fact.

Overview of the cross-border dimension of conflicts in the Great Lakes and Mano River regions

The porous nature of the borders of the Great Lakes and the Mano River regions, the artificial creation of the boundaries by the colonial masters, their improper delineation and demarcation, and the absence of an effective state control mechanism to police and patrol the borders contribute to the cross-border dimension of intrastate conflict in both regions. Similarly, the peculiarity of the border areas in terms of their proximity to each other, ethnic affinity of the people at the borderlands, the struggle for control of natural resources by rebel groups and warlords and external intervention by concerned groups and individuals add impetus to the cross-border dimension of intrastate conflicts in both regions. For instance, Filip Reyntjens (2009) reveals how the proximity of the countries in the Great Lakes region has aided the cross-border dimension of intrastate conflicts driven by several factors across the borders of one country to another. He affirms that Zaire (now the DRC) is surrounded by nine neighbouring countries, seven of which were endemically or acutely unstable and how in a perverse cycle, the instability of its neighbours threatened Zaire, just as Zaire's instability in Northern Kivu was a menace to its neighbours.

Martin Meredith (2005) in his account of the state of Africa 50 years after independence showed how the propinquity of the borders in the Great Lakes and Mano River regions of the continent aided cross-border dimensions of intrastate conflicts from Liberia to Sierra Leone as well as from Rwanda to the DRC. He revealed the cross-border support the government of Burkina Faso and Guinea gave to Charles Taylor's National Patriotic Front of Liberia (NPFL) at the outbreak of intrastate conflict in Liberia. Likewise, Charles Taylor's NPFL gave cross-border support to Revolutionary Union Front (RUF) in their invasion of Sierra Leone from the Mano River region of West Africa. In the same vein, in the Great Lakes region Meredith tells of the support the Ugandan government gave to the Rwanda Patriotic Front (RPF) across the Uganda–Rwanda border, when they marched into Rwanda in 1994.

Caitriona Dowd and Clionadh Raleigh (2015) in their analysis of the patterns of conflict and violence in Liberia and Sierra Leone from 1997 to 2003, provide insight into the cross-border dimension of intrastate conflict across the borders of both countries. They affirm that cross-border actors between Liberia and Sierra Leone included Liberians United for Reconciliation and Democracy (LURD). This rebel group in Liberia was sponsored by the Sierra Leonean government to resist Charles Taylor's authority in Liberia and the Revolutionary Union Front (RUF) based in Sierra Leone, backed by Charles Taylor to destabilise Sierra Leone and gain control of her diamond field. Both rebel groups engaged each other across the Liberia–Sierra Leone border and contributed to the dimension of intrastate conflicts in the Mano River region. Paul Richards (1996) narrates developments from Koindu in the north to the Mano River in the south of the Liberia–Sierra Leone borders, during the outbreak of intrastate conflicts in both countries. He revealed the aims and activities of the RUF, armed and equipped by Charles Taylor's NPFL to seize power in Sierra Leone and in addition, control its rain forest and diamond fields. Richards concludes that the uncontrolled character of this international border, and the history, social organisation and resource base of the communities in and around its central boundary wilderness, help explain how and why the RUF was able to secure an initial foothold, and why this dissident movement has proved so difficult to dislodge.

Intrastate conflicts in the Great Lakes and Manor River regions

The Great Lakes region

The Great Lakes region comprises countries in Central and Eastern Africa, which are members of the International Conference of the Great Lakes Region (ICGLR). They include Rwanda, Burundi, Democratic Republic of Congo (DRC), Uganda, Tanzania, Zambia, Republic of Congo, Central African Republic (CAR), South Sudan, Kenya and Sudan (Kanyangara, 2016). Rwanda, Burundi and the DRC will be the specific contexts of this study. Countries in the Great Lakes region were former colonies of Britain, Belgium and France. However, the three countries in focus were colonised by Belgium. Belgium like other colonial masters in the region left an indelible imprint with remarkable effects on the region in the post-colonial period. The outbreak of intrastate conflicts in the Great Lakes region, particularly in Rwanda and Burundi, dates back to the Cold War period (Lemarchand, 1998; Sikenyi, 2013). The post-Cold War period opened a new epoch of intrastate conflicts in the region with cross-border dimensions and regionalisation of conflicts, as it caused an alarming refugee crisis, forced migration, emergence of ethnic militia equipped to kill rival ethnic groups and external interventions with the mandate to find a lasting solution to the conflicts (Fofana, 2009; Tom, 2014; Rwengabo, 2014).

The eruption of intrastate conflicts in Africa in the post-Cold War period and its cross-border dimension, as it spilled over to neighbouring countries, is attributed to several factors. In the Great Lakes region inequitable distribution of land, ecological and environmental factors, exploitation of mineral resources, absence of internal democracy structures such as good governance, rule of law, equity, tolerance, equality and justice among others, accounted for the outbreak of intrastate conflicts in the region (Shyaka, 2008; Daley, 2006; and Polzer, 2002; Lemarchand, 1997; Bigagaza, Abong and Mukarubuga, 2002). However, in Burundi, Rwanda and the DRC entrenched ethnic rivalry and division between the Hutus and Tutsis and forced migration and consequent civil wars, have been identified as the major factors responsible for the cross-border dimension of intrastate conflicts in these countries.

Therefore, the cross-border dimension of intrastate conflicts in the Great Lakes region is due to a combination of factors, though rooted in the struggle for and against ethnic domination; with links to colonial and post-colonial developments in the region, particularly interethnic relations between the Hutus and Tutsi. Other factors include porous borders, identity questions, forced migration and displacement and external actors. Amici (1999) captures the ethnic impact of the cross-border dimension of intrastate conflicts in some states in the Great Lakes region. He states that:

> To understand the regional dimension of ethnicity in the Great Lakes region, one first has to understand that the ethnic distribution of Hutus and Tutsi is not confined within political boundaries. More than two million Hutus and Tutsis are located across the boundaries of Rwanda and Burundi in neighbouring states. Some trace their ancestry to either the DRC's north Kivu province (Banyarwanda) or its south Kivu province.
>
> (Amici, 1999: 4)

In Rwanda, Burundi and the DRC, the porous and unpoliced nature of the borders combined with ethnic rivalry, identity questions, migration issues and external factors aided the spillover of intrastate conflicts across their border in the post-Cold War era. In Burundi and Rwanda, during the 1993 and 1994 genocide, despite other factors that contributed to the killings of Hutu and Tutsis, ethnic rivalry was considered as the main factor responsible for the carnage. As already noted, the rivalry between both ethnic groups can be traced to the colonial period, and was sustained by the post-colonial governments through ethnic favouritism shown to the Tutsis against the Hutu. Events following the genocide witnessed reprisal attacks by Hutu and Tutsi militias and rebel groups against each other and against innocent civilians across their borders in Uganda and the DRC. For instance, after the 1994 genocide in Rwanda, Hutus fled en masse into the DRC provinces of north and south Kivu, where they disrupted an already-volatile situation. On 4 July 1994, as RPF forces gained control of Rwanda, for fear of revenge for their atrocities on Tutsi, Hutu genocidaires organised a mass exodus of Hutu population across of DRC. In the words of Martin Meredith:

> The roads to Zaire became choked with hundreds of thousands of Hutu, in trucks, cars, on bicycles, on foot, taking their livestock and what belongings they could carry. Buildings were stripped of window frames, door handles and corrugated-iron sheet. In two days about a million people (Hutu) cross into Zaire.
>
> (Meredith, 2005: 522)

The Hutu population from Rwanda formed alliances with DRC Hutus and their militia groups against the DRC Tutsis, and created various armed groups to defend themselves. The Hutus and Tutsis recruited combatants – Interahanwe (Hutu militia) and Rwanda Patriotic Army (RPA, Tutsi led) –

among their ethnic compatriots in Rwanda, Burundi and the DRC. This development led to the internationalisation of conflict in the Great Lakes region in the post-Cold War period (Kanyangara, 2016).

The identity or citizenship question has remained a topical issue in the Great Lakes region, particularly in Uganda and the DRC, between the indigenes of both countries and Hutu and Tutsi settlers. The governments of Uganda and the DRC for political reasons revoked the citizenship granted to the Banyarwanda and Banyamulenge settlers, some of whose ancestors had settled in the respective countries before the colonial era. The Banya-mulenge (Tutsis) in the Ugandan army, who aided President Museveni's rise to power in Uganda during the Ugandan civil war, were in return supported by Museveni in their desire to return to Rwanda and end the killings of fellow Tutsis, during and after the 1994 genocide (Mamdani, 2002). In the DRC, the Tutsis were the most affected by the decision of the Congolese government to revoke the citizenship of Hutus and Tutsis in the DRC. The Tutsis were at the mercy of Hutu militia and the indigenous Mai-Mai rebels. However, the Tutsis relied on the RPF to aid and defend them in the DRC. Kigame defended his Tutsi kin in the DRC from Hutu killings by crossing Rwandan borders into DRC. The development was a major trend in the cross-border patterns of intrastate conflicts in the Great Lakes region, in which one state violates the borders of its neighbours. Habyarimana did the same before his assassination, when he came to the aid of his Hutu brethren in south Kivu, in the DRC (Kanyangara, 2016).

Forced migration, one of the major consequences of intrastate conflicts in the Great Lakes region, played a contributory role in the spillover of conflict from Burundi, Rwanda and the DRC. Driven by ethnic sentiment, some refugees, particularly in the DRC and Rwanda, connived with rebel groups of the same ethnicity in host and receiving countries to launch attacks on rival ethnic groups within and across the borders. This development created fear and insecurity in the various refugee camps scattered across the Great Lakes (Kanyangara, 2016).

The bitter ethnic rivalry between the Hutus and Tutsis originated during the colonial period. In Burundi and Rwanda in particular, the Belgian colonial authority favoured the Tutsi minority over the Hutu majority in sharing positions of authority. The development created acrimony in their relations, and was worsened in the post-colonial period by the actions of some of the political leaders in the region. This development triggered violent retaliatory actions carried out by both ethnic groups in the post-colonial period as they sought to dominate each other through control of political power. The bitter ethnic rivalry in the post-Cold War era led to the assassination of political leaders such as Burundi's President Melchior Ndadaye of Hutu origin in 1993 and Rwanda's Hutu President Juvenal Habyarimana in 1994, senior military leaders of Hutu and Tutsi background and state sponsored killings along ethnic lines (Meredith, 2005). This climaxed in bitter ethnic rivalries, particularly the genocide carried out against Tutsis by Hutus in Burundi in 1993, of Tutsis and Hutus by Hutus in Rwanda in 1994 and of Hutus by Tutsis in the DRC in 1996–1997 (Lemarchand, 1998). Furthermore, the bitter ethnic rivalry between the Hutus and Tutsis led to forced migrations as Tutsis and Hutus fled to different refugee camps scattered across the region. Unfortunately, some of these refugee camps, especially those in north and south of Kivu in the DRC, were attacked by armed militias (Interahanwe, dominated by Hutus) and rebel groups (all-Tutsi Rwanda Patriotic Army, RPA). The deep rooted ethnic rivalry and forced migration between the Hutus and Tutsis in Burundi and Rwanda and by extension the DRC, in addition to the porous nature of the borders between these countries led to the regionalisation or internationalisation or cross-border dimension of intrastate conflicts in the Great Lakes region.

The Manor River region

The Mano River region is located in West Africa, comprising Liberia, Sierra Leone, Guinea and Cote d'Ivoire (Allouche, Benson and M'Cormack, 2016: 4). Interestingly, all the countries in the Mano River region have witnessed intrastate conflicts in the post-Cold War period (Adedeji, 1999). Liberia and Sierra Leone will be the focus of this study. The causes or factors that led to the outbreak of intrastate conflicts in Liberia and Sierra Leone are similar to those of other countries that experienced intrastate conflicts in post-Cold War West Africa. Some of these include corruption, ethnic marginalisation, tenure elongation, poor governance, social injustice, breakdown of democratic institutions, external interference and quest for wealth by rebel leaders (Sawyer, 2004; Stanley, 2004; Annan, 2014).

Of all these factors ethnic marginalisation, quest for political control, external interference and control of economic resources by political leaders and by rebel groups not only fuelled the intrastate conflicts in Liberia and Sierra Leone, but also contributed to the spillover of the conflicts across the borders of both countries. Ethnic marginalisation and struggle for political control in Liberia and Sierra Leone can be linked to their unique history of being settler colonies of returning and repatriated and rescued slaves in the nineteenth century (Onwubiko, 1973). For instance, in Liberia, resettled slaves from the USA known as Americo-Liberians dominated the country politically from independence in 1847 until 1980, when political power was seized by the indigenes through a military coup (Adeleke, 1995). Sierra Leone was settled by returnee slaves from Nova Scotia in Canada and later recaptured slaves after the abolition of the slave trade in 1807. The recaptured slaves included people from many nations, with the majority being the Yoruba, the Ibo and the Asante (Onwubiko, 1973). They were called the Creoles. Like the Americo-Liberians, the Creoles were educated and dominated the political and economic affairs of their colonies.

Still in Mano River region, the struggle for ethnic supremacy was manifested in the political oppression of one ethnic group by another. In Liberia and Sierra Leone, the Americo-Liberians and the Creoles carried out deliberate plans to deprive the indigenes' access to power. Unfortunately, their hold on power provoked reactions from the indigenes and this contributed to the outbreak of civil wars in both countries. According to Stanley (2004) neither the Krio (Creoles) nor the Americo-Liberians, both of whom possessed considerable technological or educational advantages over the peoples of the interior, were able to sustain their privileged positions over a period of geopolitical change.

Furthermore, following overthrew of the Americo-Liberians from power by the indigenous Liberians in 1980, there emerged a struggle between the different indigenous ethnic groups for political power in Liberian politics. According to Godwin Ndubuisi Onuoha:

> Samuel Doe's administration epitomized horizontal struggles between the different ethnic groups for power and control. Doe was of the Krahn ethnic extraction, for this reason the Krahn became the privileged, while the Gio, Mano were relegated to the background. In a progressive and systematic manner, Doe gradually eliminated the Gio, Mano and Americo-Liberians from his government for the perceived fear of their disloyalty and dissent ambitions.
>
> (Onuoha, 2004: 27)

On the other hand, in Sierra Leone, weak political institutions and rivalry among its leaders, particularly the military leaders and rebel groups, partly contributed to the outbreak of civil war in the country from 1991 to 2002 (Stanley, 2004). Foday Sankoh's Revolutionary United Front (RUF) rebel group was supported by Charles Taylor (National Patriotic Front of Liberia) to oust the civilian administration of President Momoh in 1991. From 1991 to 2000, Sierra Leone was engulfed in civil wars fuelled by power tussles among its military leaders and rebel groups struggling partly to control the country's diamond fields.

Like in the Great Lakes region, therefore, several factors contribute to the cross-border dimension of intrastate conflicts Liberia and Sierra Leone. One of these is international collaborations between different actors located within and beyond the states involved in intrastate conflicts. For example, Charles Taylor, the rebel leader of the NLPF, and Fondah Sankoh, leader of the RUF in Sierra Leone were supported, trained and financed in their war enterprises by Colonel Muammar Gaddafi of Libya, Blaise Compaoré of Burkina Faso and Houghouet-Boigny of Cote d'Ivoire to overthrow the respective governments of their countries and likewise create instability in the region (Ellis, 2001; Richards, 1996; Allouche et al., 2016). Blaise Compaore and Houghouet-Boigny were bitter about Samuel Doe, for killing relatives of theirs when he seized power in 1980 (Meredith, 2005). This development resulted in the transnationalisation and regionalisation of the Mano River conflicts involving different groups with vested interests in the conflicts. The Liberian civil war started in 1989, following the Charles Taylor-led NLPF attack from the borders of Burkina Faso to topple the government of Samuel Doe. In 1991, Charles Taylor provided support for Fondah Sankoh's RUF to create instability in Sierra Leone, leading to the Sierra Leonean civil war. The entering of the RUF into Sierra Leone with the support of Charles Taylor's rebel group and the leader of Burkina Faso dissidents destabilised Sierra Leone through the outbreak of the Sierra Leone civil war. The development opened a new phase of a brutal intrastate conflict in the country that witnessed the amputation and killings of men, women and children (*The Guardian*, 2003).

The quest for wealth by various warlords contributed to the spillover or cross-border dimension of the Mano River region conflicts in Liberia and Sierra Leone in the post-Cold War era. The interests of factional groups were characterised by a synergistic relationship between political, military and economic calculations – economic opportunities became an important factor in reproducing conflict and undermining the prospect of peace (Aning, 2003: 100). Both Sierra Leone and Liberia are countries rich in mineral deposits such as diamonds, rubber, iron ore and timber. Charles Taylor's support for Fondah Sankoh's RUF was to enable him have access to the diamond mines in the east of the country, which would not only profit both parties, but provide the resources with which to purchase arms to prolong the war. It equally led to Charles Taylor being a major beneficiary of exploitation of Sierra Leonean diamonds during the civil war (Stanley, 2004). Similarly, Charles Taylor's control of parts of Liberia enabled him to have access to some of the country's natural resources such as the Liberian timber, which he sold on the international market (Marc et al., 2015).

The recruitment of soldiers cut across the borders of some of the countries in the Mano River region. With respect to the outbreak of the Sierra Leonean and Liberian civil wars, some of the rebel fighters that helped Charles Taylor and Foday Sankoh to launch their attacks came from outside their borders. For instance, Charles Taylor had mercenaries who came from Burkina Faso, while a large part of the core of Foday Sankoh's invading RUF was drawn from Taylor's "special forces" (Sawyer, 2004). Interestingly, the Sierra Leone government, in an attempt to repel the attacks of the RUF forces in Freetown its capital, sourced the services of Liberian refugees to supplement the efforts of its already demoralised army. Many of these had been soldiers in the Liberian army. It was this group that became the United Liberation Movement for Democracy (ULIMO), and launched a cross-border assault on the NPFL in August 1991.

Resolving intrastate conflicts that have cross-border dimensions in the Great Lakes and Mano River regions

The Great Lakes region

The leaders of Burundi, Rwanda and the DRC contributed to the intrastate conflicts spilling from one country to another along ethnic lines. For instance, before the assassination of President Juvenal Habyarimana of Rwanda and the 1994 genocide in his country, he promoted Hutu nationalist consciousness beyond the borders of Rwanda. Hutu nationalist consciousness was an ideology held by some Hutu extremists in Habyarimana's administration. Their rhetoric was based on both a sense of victimisation from Tutsi domination during the colonial times, as well as of a Tutsi rebellion formed in Uganda led by the RPF (Fofana, 2009; Mamdani, 2002). Therefore, this extremism, led by Habyarimana was committed to protect and defend the Hutu population whether they were of Rwandan extraction or not. This explains Habyarimana-led intervention on behalf of Hutu in south Kivu in the DRC, following eruption of land conflicts between Congolese Hutu and the indigenous authority in south Kivu. He meddled with the domestic affairs of the DRC's north Kivu region, between the Hutus and the indigenes. In the same vein, Paul Kigame, who led the Rwanda Patriotic Front (RPF), likewise intervened in south Kivu to avert the killings and victimisation of Tutsis by Hutus soldiers and the Congolese government, under Joseph Kabila (Fofana, 2009). The Kigame-led intervention in the DRC on behalf of his kin across the Rwandan border was based on the premise of eliminating the Hutu soldiers in south Kivu because they were preparing to launch an assault in Rwanda. Similarly, Paul Kigame and the RPF saw themselves as liberators of Tutsi people across the Great Lakes region (Fofana, 2009: 35).

The humanitarian and material catastrophes that emanated from intrastate conflicts in the Great Lakes region and its cross-border dimension galvanised several attempts by leaders and intergovernmental organisations in the region to find a lasting solution to the carnage. Similarly, the United Nations (UN) and the European Union (EU), as well as former colonial masters of the region have been involved in finding a lasting solution to conflicts in the affected states (Meredith, 2005). In the process of searching for a solution, several peace agreements spearheaded by the Organisation of African Unity (OAU), now African Union (AU) were reached, namely the Arusha Peace on Rwanda (1993), on Burundi (2000) and on the DRC (2000). The AU spearheaded peacekeeping missions in the form of the Neutral Military Observer Group (NMOG) in Rwanda, and African Mission in Burundi (AMIB) with troops from South Africa, Ethiopia and Mozambique and OAU/UN missions in Congo before the UN's mission, MUNOC (Daley, 2006). The African peacekeeping efforts in the Great Lakes region thus preceded the UN-led peacekeeping missions.

However, it is imperative to state that Africa-led peacekeeping missions are faced with some challenges such as funding and logistical support which to some extent mar their operations and mission. This has been the situation in the Great Lakes region. Because of this, as in other parts of the continent confronted with threats of intrastate conflicts, the UN took over the missions, but was itself hampered by Security Council decisions, restrictive mandate and lack of cooperation from the local people (Sikenyi, 2013).

The peace agreements reached in ending intrastate conflicts and its cross-border dimensions have remained fragile due to several factors: lack of political will of the parties involved; division among political leaders in the region; and citizens' lack of faith on the peace process. Campbell (2002) argues that lasting peace in the affected states confronted by intrastate conflicts, particularly the DRC, is hindered by factionalism within the rebel movements and militarisation of local communities, coupled with poor infrastructure and reduced state effectiveness in policing territory. Intrastate conflict and its cross-border dimension have reduced significantly in Burundi and Rwanda in recent times; however, there are still pockets of attacks in eastern Congo as well as north and south Kivu, because of the large numbers of Hutu and Tutsi settlers in the province.

The Mano River region

The leaders of the West African sub-region through various efforts made attempts to avert the spread of intrastate conflicts in the Mano River region, particularly to avoid spillover effects from Liberia and Sierra Leone into other neighbouring countries. Several mediation efforts were embarked upon by elder statesmen within and outside the affected states, religious bodies and non-governmental organisations to persuade the rebel groups and the government of Liberia and Sierra Leone to sheath their swords and end the civil wars for the sake of their brothers and sisters (Gberie, 2006). Similarly, the following agreements have been reached to end the conflicts in Liberia and Sierra Leone: the Abuja Accord of 1995, Abuja Accord II of 1996 and the Peace Agreement of 2002 signed in Burkina Faso's capital Ouagadougou that ended the Sierra Leone civil war (Riley and Sesay, 1996: 68). The leaders of the West African sub-region, led by Nigeria, in 1990 established the ECOWAS Monitoring Group (ECOMOG). ECOMOG was mandated by ECOWAS member states to carry out peacekeeping missions in Liberia and Sierra Leone and other countries in the sub-region threatened by civil war. However, intrastate conflict and its spillover effect in the Mano River of West Africa ended following Charles Taylor's decision to seek political asylum in Nigeria in 2003.

With regard to peaceful and reconciliatory efforts to end intrastate conflict and its cross-border dimension in both the Great Lakes and the Mano River regions, the sub-regional organisations in both regions, such as the International Conference on the Great Lakes Region (ICGLR), the Mano River Union (MRU) and the Economic Community of West African States (ECOWAS) have initiated several efforts towards this direction. In the Mano River region, peace has gradually returned since Charles Taylor left power in 2003. In the Great Lakes region, peace is still fragile, particularly in the north and south Kivu, because of recurrent attacks by ethnic militias and rebel groups along the borders. Peace is central in finding a lasting solution to the cross-border dimension of intrastate conflicts in the Great Lakes and Mano River regions of Africa. The leaders of the Great Lakes region and the states involved in conflicts must consciously work towards the return of peace in the region.

Conclusion

The chapter has examined the cross-border dimension of conflicts in the Great Lakes and Mano River regions in the post-Cold War period. It has shown the various factors that led to the outbreak of intrastate conflicts as well as their cross-border dimensions in Burundi, Rwanda, the DRC, Liberia and Sierra Leone. Similarly, the various efforts made by the leaders of the regions and other groups to resolve the conflicts were examined, and the extent to which they have succeeded in line with the goals of mediation. The study reveals that the porous nature of the borders in the respective regions contributed to the cross-border dimension of these intrastate conflicts.

Given the above, the following recommendations can be made. First, greater importance should be placed on border security by the leaders of both regions. In addition, credible and effective border management institutions and policies to address their border problematic should be established at the local and sub-regional levels. ECOWAS has made some giant strides in this regard through the ECOWAS Conflict Framework established in 2008 (ECOWAS, 2008). Section VIII on Cross-Border Initiatives of the ECOWAS Conflict Prevention Framework outlines measures to address conflicts across the borders of member countries. Second, countries in the Great Lakes and Mano River regions should endeavour, in line with the AU Border Program (AUBP), to ensure that their borders are clearly demarcated (Asiwaju, 2015). This is critical in order to prevent any group from illegally encroaching into another country's territory. Similarly, a well demarcated border would ensure effective border policing and management. Although African borders have remained porous, proper management in terms of border security would help address cross-border dimensions of intrastate conflicts.

References

Adedeji, Adebayo. 1999). Comprehending African Conflict. In Adebayo Adedeji (Ed.), *Comprehending and Mastering African Conflicts: The Search for Sustainable Peace and Good Governance.* London: Zed Books.

Adeleke, Ademola. 1995. The Politics and Diplomacy of Peacekeeping in West Africa: The ECOWAS Operation in Liberia. *The Journal of Modern African Studies* 33(4): 569–593.

Akokpari, K. John. 1999. The Political Economy of Migration in Sub-Saharan Africa. *African Sociological Review* 3(1): 75–93.

Allouche, Jeremy, Benson, Matthew and M'Cormack, Freida. 2016. *Beyond Borders: The End of the Mano River War(s)?*London: Institute of Development Studies.

Amici, R. 1999. *Conflict Resolution in the Great Lakes Region.* Conference Report. London: The Council of the Democratic Federal Republic of Congo.

Anene, J. C. 1970. *The International Boundaries of Nigeria 1885–1960: The Framework of an Emergent African Nation.* London: Longman.

Aning, Kwesi Emmanuel. 2003. Regulating Illicit Trade in Natural Resources: The Role of Regional Actors in West Africa. *Review of African Political Economy* 30(95): 99–107.

Aning, Kwesi and Pokoo, John. 2017. Between Conflict and Integration: Border Governance in Africa in Times of Migration . *International Reports* 1: 54–65.

Annan, Nancy. 2014. Violent Conflicts and Civil Strife in West Africa: Causes, Challenges and Prospects . *International Journal of Security & Development* 3(1), Art. 3. DOI: doi:10.5334/sta.da

Asiwaju, A. I. 1984. Artificial Boundaries. An Inaugural Lecture delivered at the University of Lagos. Lagos: University of Lagos Press.

Asiwaju, A. I. 2003. *Boundaries and African Integration: Essay in Comparative History and Policy Analysis.* Lagos: Panaf Publishing.

Asiwaju, A. I. 2015. Borders in African Policy History. In A. I. Asiwaju (Ed.), *Borders in Africa: An Anthology of the Policy History.* Addis Ababa: Institute for Peace and Security Studies.

Azam, Jean-Paul. 2001. The Redistributive State and Conflicts in Africa. *Journal of Peace Research* 38(4): 429–444.

Azar, E. 1990. *The Management of Proctracted Social Conflict: Theory and Cases.* Aldershot: Dartmouth.

Bigagaza, Jean, Abong, Carolyn and Mukarubuga, Cecile. 2002. Land Scarcity, Distribution and Conflict in Rwanda. In Jeremy Lind and Kathryn Stuman (Eds.), *Scarcity and Surfeit: The Economy of Africa's Conflict.* Pretoria: Institute for Security Studies.

Cameron, Fraser. 2005. *U.S Foreign Policy after the Cold War: Global Hegemon or Reluctant Sheriff?* (2nd edition). London: Routledge.

Campbell, H. 2002. *The Lusaka Peace Agreement: The Responsibility of African Academics and Civil Society.* Nairobi: Nairobi Peace Initiative.

Cilliers, Jakkie and Schünemann, Julia. 2013. *The Future of Intrastate Conflict in Africa: More Violence or Greater Peace?*Pretoria: Institute for Security Studies.

Collier, Paul and Hoeffler, Anke. 2004. Greed and Grievance in Civil War. *Oxford Economic Papers* 56: 563–594.

Daley, Patricia. 2006. Challenges to Peace: Conflict Resolution in the Great Lakes Region of Africa. *Third World Quarterly* 27(2): 303–319.

Djelic, Marie-Laure and Quack, Sigrid. 2010. *Transnational Communities: Shaping Global Economic Governance.* Cambridge: Cambridge University Press.

Dowd, Caitriona and Raleigh, Clionadh. 2015. Mapping Conflict across Liberia and Sierra Leone. In Consolidating Peace: Liberia and Sierra Leone. *Accord* 23: 13–18.

ECOWAS. 2008. *ECOWAS Conflict Prevention Framework.* responsibilitytoprotect. org/ECOWAS%20ECPF%20Ekiyor.pdf (accessed March 2019).

Ellis, S. 2001. *The Mask of Anarchy: The Destruction of Liberia and the Religious Dimension of an African Civil War.* New York: NYU Press.

Ewald, Jonas. 2004. *A Strategic Conflict Analysis for the Great Lakes Region.* Göteborg University: SIDA.

Faist, Thomas. 2009. Diaspora and Transnationalism: What Kind of Dance Partners? In Rainer Bauböck and Thomas Faist (Eds.) *Diaspora and Transnationalism: Concepts, Theories and Methods.* Amsterdam: IMISCOE – University Press Series.

Fofana, Idriss. 2009. A Crisis of Belonging: Rwanda's Ethnic Nationalism and the Kivu Conflict. *Harvard International Review* 30(4): 34–38.

Gberie, Lansana. 2006. Bringing Peace to West Africa: Liberia and Sierra Leone. Africa Mediators' Retreat.

Griffiths, Ieuan. 1986. The Scramble for Africa: Inherited Political Boundaries. *The Geographical Journal* 152(2): 204–216.

*The Guardian*2003. Former Sierra Leonean Rebel Leader Sankoh Dies. www.thegua rdian.com/world/2003/jul/30/westafrica.sierraleone (accessed March 2019).

Gurr, Ted Robert. 2009. Peoples against States: Ethnopolitical Conflict and the Changing World System – 1994 Presidential Address. In Rajat Ganguly (Ed.), *Ethnic Conflict: Causes of Ethnic Conflict* (Volume II). London: Sage.

Hargreaves, J. D. 1984. The Making of the Boundaries: Focus on West Africa. In A. I. Asiwaju (Ed.), *Partitioned Africans: Ethnic Relations across Africa's International Boundaries 1884–1994*. London: C. Hurst & Company.

Hauss, Charles. 2010. *International Conflict Resolution*. New York: Continuum International.

Hopmann, P. Terrence. 1999. *Building Security in Post-Cold War Eurasia: The OSCE and U.S. Foreign Policy*. Washington, DC: United States Institute of Peace.

Ikome, Francis Nguendi. 2012. *Africa's International Borders as Potential Sources of Conflict and Future Threats to Peace and Security*. Pretoria: Institute for Security Studies.

Kanyangara, Patrick. 2016. Conflict in the Great Lakes Region: Root Causes, Dynamics and Effects. *Conflict Trends* 1(January): 3–11.

Lemarchand, René. 1997. Patterns of State Collapse and Reconstruction in Central Africa: Reflections on the Crisis in the Great Lakes Region. *Africa Spectrum* 32(2): 173–193.

Lemarchand, René. 1998. Genocide in the Great Lakes: Which Genocide? Whose Genocide? *African Studies Review* 41(1): 3–16.

Malejacq, Romain. 2007. Looking at the Individual in Liberia and Sierra Leone: From a Regional Conflict to a "Human Insecurity Complex". Revue de la Sécurité Humaine / *Human Security Journal* 3: 43–54.

Mamdani, Mahmood. 2002. African States, Citizenship and War: A Case-Study. *International Affairs* 78(3): 493–506.

Mamdani, Mahmood. 2005. Political, Identity, Citizenship and Ethnicity in Post-Colonial Africa. Keynote address at the Arusha Conference, "New Frontiers of Social Policy" 12–15 December.

Marc, Alexandre, Verjee, Neelam and Mogaka, Stephen. 2015. *Responding to the Challenge of Fragility and Security in West Africa*. Washington, DC: World Bank.

McWilliam, C. Wayne and Piotrowski, Harry. 1997. *The World Since 1945: A History of International Relations*. London: Lynne Rienner Publishers.

Meredith, Martin. 2005. *The State of Africa: A History of Fifty Years of Independence*. London: Simon & Schuster.

Nhema, Alfred. 2008. Introduction: The Resolution of African Conflict. In Alfred Nhema and Paul Tiyambe Zeleza (Eds), *The Resolution of African Conflicts: The Management of Conflict Resolution and Post-Conflict Resolution*. Athens: Ohio University Press.

Oketch, Johnstone Summit, and Polzer, Tara. 2002. Conflict and Coffee in Burundi. In Jeremy Lind and Kathryn Stuman (Eds.), *Scarcity and Surfeit: The Economy of Africa's Conflict*. Pretoria: Institute for Security Studies.

Olayode, Kehinde. 2016. Beyond Intractability: Ethnic Identity and Political Conflicts in Africa. *International Journal of Humanities and Social Science* 6(6) (June): 242–248.

Onuoha, Godwin Ndubisi. 2004. Local and External Intersections in African Conflicts: Trends and Perspectives in the Liberian Experience. *Nigerian Journal of Policy and Development* 3: 23–30.

Onwubiko, K. B. C. 1973. *School Certificate History of West Africa Book Two: 1800–Present Day.* Onitsha, Nigeria: Africana Educational Publishers.

Rajaee, Farhang. 2000. *Globalisation on Trial: The Human Condition and the Information Civilisation.* Ottawa: International Development Research Centre.

Ramsbotham, Oliver, Woodhouse, Tom and Miall, Hugh (Eds). 2005. *Contemporary Conflict Resolution: The Prevention, Management and Transformation of Deadly Conflict* (2nd ed.). Cambridge: Polity Press.

Reyntjens, Filip. 2009. *The Great African War: Congo and Regional Geopolitics, 1996–2006.* Cambridge: Cambridge University Press.

Richards, Paul. 1996. The Sierra Leone–Liberia Boundary Wilderness: Rain Forest, Diamond and War. In Paul Nugent and A. I. Asiwaju (Eds.), *African Boundaries: Barriers, Conduits and Opportunities.* London: Pinter.

Riley, Stephen and Sesay, Max. 1996. Liberia: After Abuja. *Review of African Political Economy* 23(69): 429–437.

Rwengabo, Sabastiano. 2014. The Migration–Interstate Conflict Nexus. *Social Affairs: A Journal for the Social Sciences* 1(1): 52–82.

Sadowski, Yahya. 1998. Ethnic Conflict. *Foreign Policy* 111: 12–23.

Sawyer, Amos. 2004. Violent Conflicts and Governance Challenges in West Africa: The Case of the Mano River Basin Area. *The Journal of Modern African Studies* 42 (3): 437–463.

Shyaka, Anastase. 2008. Understanding the Conflicts in the Great Lakes Region: An Overview. *Journal of African Conflict and Peace Studies* 1: 5–12.

Sikenyi, Maurice. 2013. Challenges of Regional Peacebuilding: A Case of the Great Lakes Region. In *Beyond Intractability*, Kroc Institute of International Peace Studies (June). www.beyondintractability.org/casestudy/sikenyi-great-lakes

Sindayigaya, Aimé. 2014. The Eastern Democratic Republic of Congo Conflict: A Long Running Fiasco and its Lessons for Regional Policy Makers. *Jambonews.net.* www.jambonews.net/en/news/20140604-the-eastern-democratic-republic-of-congo-conflict-a-long-running-fiasco-and-its-lessons-for-regional-policy-makers/

Stanley, William R. 2004. Background to the Liberia and Sierra Leone Implosions. *GeoJournal* 61(1): 69–78.

Themner, Lotta and Wallensteen, Peter. 2012. Patterns of Organised Violence, 2001–2010. In *SIPRI Yearbook 2012: Armaments, Disarmament and International Security.* Oxford: Oxford University Press.

Tom, Ogwang. 2014. Armed Conflicts and Forced Migration in the Great Lakes Region of Africa: Causes and Consequences. *International Journal of Research in Social Sciences* 4(2): 147–161.

UNESCO 2013. Trans-nationalism: Social and Human Science – International Migration. www.UNESCO.org/new/en/social-and-human-science/themes/international-migration (accessed 12 October 2018).

van Schendel, Willem and Abraham, Itty (Eds). 2005. *Illicit Flows and Criminal Things: States, Borders and the Other Side of Globalisation.* Bloomington: Indiana University Press.

Wallensteen, Peter and Sollenberg, Margareta. 2001. Armed Conflict, 1989–2000. *Journal of Peace Research* 38(5): 629–644.

Walt, M.Stephen. 2005. The Relationship between Theory and Policy in International Relations. *Annual Review of Political Science* 8(1): 23–48.

Wilson, Thomas M. and Donnan, Hastings (Eds). 2012. *A Companion to Border Studies.* Oxford: Blackwell Publishing.

Zartman, I. William. 1965. The Politics of Boundaries in North and West Africa. *The Journal of Modern African Studies* 3(2): 155–173.

Zartman, I. William. 2010. Identity, Movement and Response. In I. William Zartman (Ed.), *Understanding Life in the Borderlands: Boundaries in Depth and in Motion.* Athens: The University of Georgia Press.

Zeleza, Paul Tiyambe. 2008. The Causes and Costs of War in Africa from Liberation Struggles to the War on Terror. In Alfred Nhema and Paul Timambe Zeleza (Eds), *The Roots of African Conflicts: The Causes and Costs.* Athens: Ohio University Press.

7 Taking sustainable development to the limits

State policy – grassroots actor activities interface in borderlands

Christopher Changwe Nshimbi

Introduction

Zambia and Zimbabwe share three notable characteristics in relation to their common border. Firstly, both are former British colonies. Secondly, they (together with present day Malawi) constituted an amalgamated Federation of Rhodesia and Nyasaland from 1953 to 1963. Thirdly, and related to the focus of this chapter, the Southern African neighbours are physically separated, in parts, by a natural border – the Zambezi River. The political boundary between them however was demarcated by the British South African Company (BSAC) when both were still called Rhodesia. After demarcation, the BSAC officially named the territory to the north of the Zambezi River, Northern Rhodesia and the one south, Southern Rhodesia. In 1964, the northern territory attained political independence and assumed the name Zambia. Southern Rhodesia then generally became Rhodesia until 1980, when it attained independence and was subsequently renamed Zimbabwe. This chapter focuses on the human activities which cross over and occur around a physical feature on the natural border that bisects Zambia and Zimbabwe. The feature in question is Lake Kariba, the world's largest man-made lake.

The concrete double-curvature arc that walls the Kariba lake or reservoir was built across the Middle Zambezi River over a four-year period (from 1955 to 1959), during the Federation of Rhodesia and Nyasaland. It is located approximately 420 km downstream from the Victoria Falls. The dam rises up to approximately 128 m, has a 617 m crest and a water-holding capacity of about 181 billion m^3. The water reservoir behind the dam stretches to a full length of about 280 km and a full width of 32 km. It squares up to a total surface area of approximately 5,400 km^2 and has a 663,000 km^2 catchment area. The reservoir thrusts water down to two underground hydroelectricity power stations: the Kariba North Bank, which has an installed capacity of 1,080 MW, and the Kariba South Bank with an installed capacity of 750 MW.

The power stations supply electricity to Zambia and Zimbabwe (and their neighbours). They are operated by the Zambia Electricity Supply Corporation (ZESCO) and Zimbabwe Electricity Supply Authority (ZESA), respectively. The Kariba hydroelectricity complex is managed by the Zambezi River

Authority (ZRA), which is jointly owned by the governments of Zambia and Zimbabwe as established by Acts of Parliament that were simultaneously passed in both countries in 1987. The ZRA, however, represents a history of cross-border cooperation that dates back to 1951, during colonialism.[1] In fact, the Kariba Dam project represents more than cross-border cooperation between two countries, because it was not only initiated by the Rhodesian colonial government for the two Rhodesias, it was also implemented in conjunction with the World Bank (Cliggett et al., 2007).

Lake Kariba is also a major location of Zambia's and Zimbabwe's fishing industries. Actually, it is central to not only Zambia's and Zimbabwe's development and energy needs but impacts those of neighbouring countries too. Notably, Botswana, Malawi, Mozambique and South Africa, though (except for Malawi) their energy needs were not part of the original motivation for building the dam. Instead, the dam was constructed to satisfy the Federation of Rhodesia and Nyasaland's need for massive energy, and its development agenda. This also fell within the strategic joint management and administration of facilities and basic services – such as education and communication – in the Federation. A unified economic and political approach to managing federal affairs made it easier for the British colonial authority to administer the federal region and created an attractive economic space for investment (Nshimbi, 2017). But Groves (2013) argues that the majority of Africans in these territories were opposed to the creation of the Federation of Rhodesia and Nyasaland, because they saw it as a ploy to extend white settler domination northwards across the Zambezi. This anti-federation sentiment, Groves posits, helped unite the political interests of Africans and raised Pan-African or regional consciousness, which peaked around the time of the All Africa People's Conference in Accra, Ghana, in 1958. This, coupled with the imbalances in the power sharing mechanisms designed by the colonialists in Rhodesia, led to the collapse of the federation in 1963.

Lake Kariba: physical, environmental, socioeconomic and cultural environs

The collapse of the federation and subsequent independence of Zambia effectively turned Lake Kariba into a sort of transboundary frontier; a frontier separating but also shared by communities whose sociocultural ties had hitherto only been bridged by the river Zambezi. The impact of the creation of the lake is evident in the socioeconomic and cultural consequences it has had on those displaced from the Zambezi or Gwembe Valley, as the rising waters flooded their villages. This is beside the impacts on the environment and wildlife. About 6,000 animals threatened and trapped on islands by the dam's rising waters had to be brought to safety in a three-year (1958–1961) rescue operation called "Operation Noah". On their part, the involuntarily displaced human inhabitants of the Valley were resettled on upland between 1956 and 1959. About 86,000 people – 55,000 on the Northern Rhodesia and

31,000 on the Southern Rhodesia side of the Zambezi River – had lived in its valley by the mid-1950s (Colson, 1960: 6). The majority included a culturally distinct ethnic group called the Gwembe Valley Tonga. They had long inhabited the Valley and engaged in subsistence agriculture in villages clustered around fertile alluvial deposits on the banks of the Zambezi (Colson, 1960; Scudder, 1962) and its tributaries. Most of them farmed on the banks of the Zambezi. They harvested two main crops a year and complemented it with, *inter alia*, fishing and hunting. They also drew their water from the Zambezi and its tributaries. The Gwembe Valley Tonga were thus self-sufficient before they were forcefully relocated. Besides economic and livelihoods effects, the relocation also impacted the sociocultural lives of Gwembe Valley Tonga. A salient impact relates to gender and women's access to land. The Gwembe Tonga practice a matrilineal culture. Consequently, access rights to the alluvial land on the Zambezi River banks before relocation were held in corporate matrilineages (Cliggett, 2002). But, according to Cliggett (2002: 213–214), the relocation brought with it imbalance in men's and women's access to land, as people had to now depend more on cleared fields rather than the alluvial soils on the Zambezi River banks. Having moved to higher ground, men had to clear bush areas to create larger fields which they ploughed using oxen (ibid.). Cliggett says that rather than clear the bush and plant such large areas themselves, most women would, now, depend on their husbands for ploughs and fields in the new location. Approximately 57,000 Gwembe Valley Tonga on both the Northern Rhodesian and Southern Rhodesian (Zambian and Zimbabwean) sides of the Zambezi were relocated due to the construction of the Kariba dam (Colson, 1960: 7; Cliggett, 2001: 5).

In the 10-year period following relocation of the Tonga, schools, cooperatives, agricultural production facilities – especially support for cash crops like cotton and sorghum – and commercial fisheries were established for the resettled communities. Commercial stores and a coal mine were also opened on the Zambian side in the late 1960s (Scudder, 1968). Some of these developments, however, were associated with what Clark et al. (1995: 95–96) call a "Period of Prosperity". It typically defined the socioeconomic environment in Zambia from 1963 to 1973. And being associated with political independence, it was characterised by post-independence economic growth and rising living standards. The Zambian government then expanded and delivered educational, agricultural and health services and physical infrastructure in the area. Also noteworthy is that during this period, Southern Rhodesia was still under colonial rule and the indigenous people there, engaged in the war for political liberation. As Zambia supported the cause, the Gwembe Valley and district were destabilised.

Besides this war, pre-flood Gwembe Valley inhabitants have not fully restored or secured their livelihoods from the time they were displaced. The involuntary relocation drastically changed their lives, undermining livelihoods and resource bases. Their system of agriculture, for instance, is said to have "changed almost overnight" after they were relocated, with almost 60,000 people changing from the intensive agriculture they previously practised on alluvial soils to extensively farming on land cleared from bushes (Cliggett, 2002: 214) on the highlands. Their life has been stressful since they were relocated. Clark et al. (1995: 95) call the time they experienced the involuntary displacement and relocation the "Period of Stress". And this was to the extent that Scudder (1968: 171) says higher adult mortality rates were recorded among them in the years immediately after relocation. Equally so, morbidity and mortality rates among children increased, despite the improvements in and provision of medical services by the states.

At some point, according to Gadd, Nixon et al. (1962: 495 as cited in Scudder, 2005: 41), women and children, but not men, died of "an acute condition of sudden onset and high and rapid mortality". According to Scudder (1968), it took the Tonga about two years after the forced relocation to revert to their former subsistence economy and adequately support the population. In some cases, it even took four years or longer to adjust.

Over the same period, the relocated people experienced insufficient harvests because there were droughts in the Valley. Drought conditions continued with increasing frequency and duration from the 1970s onwards and negatively affected agricultural production. Because of this, governments had to provide famine relief. To date, the inhabitants of the Valley face famine and rely on government and non-governmental organisation (NGO) relief services (see, e. g., World Vision, 2003; Xinhua, 2008; UN Development Programme, 2017).

Economically, the downturn in the Zambian economy from the mid-1970s eroded whatever improvements had accompanied the agricultural input credit facilities, health and transport networks and infrastructure and consumer goods that had been seen in the Valley in the immediate post-independence period. For Zambia, the economic decline was linked to the global decline in oil prices that occurred in 1973. This led to a fall in the global price of copper, which to date is the mainstay of Zambia's economy. Clark et al. (1995: 96) call the economic environment that characterises the Valley (and indeed Zambia in general) during this downturn, a "Period of Decline". Since the 1980s, the Valley has experienced the same economic and social conditions which are generally prevalent in Zambia and Zimbabwe today. The conditions can be characterised as declining standards of living, poor service delivery and high levels of unemployment. They are a consequence of the austerity period brought about by structural adjustment programmes (SAPs) instituted by the neighbouring governments in the late 1980s (for Zambia) and early 1990s under the auspices of the International Monetary Fund (IMF) and World Bank. The SAPs have contributed to, *inter alia*, further marginalisation of the Valley Tonga, who already lived in a physical and socioeconomic environment characterised by extreme climatic conditions and volatile political economy (Cliggett et al., 2007). For, extreme climatic conditions manifested as severe droughts and flooding, and pest infestation as well as the environmental and socioeconomic effects of constructing the Kariba Dam have affected the lives of the Gwembe Tonga for decades.

Besides this, the cultural consequences of the flood associated with constructing the dam are best told by Scudder (1968: 171–172):

> During this prolonged period of stress a number of behavioral patterns were temporarily or permanently stopped. This was especially the case with the 6000 Tonga who, because of inadequate farm land near, or inland from, the land shore were relocated in the Lusitu area below the dam site. Unlike other evacuees who were shifted relatively short distances, those bound for the Lusitu were moved approximately 100 miles to a relatively unknown area with a bad reputation. [...] Both before and after relocation the Tonga lived in neighborhoods (cisi) composed of a number of village communities, the distribution and density of population being closely correlated with the availability and fertility of agricultural soils. Formerly each cisi had one or more neighborhood ritual leaders (sikatongo) whose prime task was the ritualized initiation of a number of important activities connected with the annual cycle. Of these, agricultural tasks (planting, bird scaring, harvesting, etc.) were among the most important, the sikatongo having his own ritual field. Following relocation the activities of sikatongo stopped entirely and to date partial attempts to reestablish them appear to be restricted to a small minority of neighborhoods outside of the Lusitu [...]. According to those Tonga involved, these have been curtailed because the neighborhood ritual leader operated within a defined territory (katongo) which has now been flooded. His prerogatives remain there. As for attempts to shift them to the relocation areas, these would be dangerous because of the opposition and retaliation of the spirits already associated with the land.

Such were the social and cultural changes induced by the flood caused by the construction of the Kariba dam. Scudder further says relocated populations suffered loss of identity. So, a complex of environmental, demographic, epidemiological and psychological factors seem to compound the socioeconomic and political conditions experienced by the resettled people in their new environs.

State power and scraping a living in Gwembe Valley: making sense of harsh realities at grassroots

The foregoing is an attempt to briefly highlight a history and some of the physical, environmental, socioeconomic and cultural conditions in which the inhabitants of the Zambezi or Gwembe Valley strive to survive by exploiting opportunities on and around the man-made waterbody on which also lies the international line that demarcates Zambia and Zimbabwe. Some Valley dwellers do this by operating in the fisheries sector. This chapter takes special interest in artisanal fishers – the grassroots or local actors who opt to take up fishing as an occupation to sustain livelihoods.

Artisanal fishing is different from the capital intensive and highly mechanised offshore industrial fishing on Lake Kariba which specialises in *Limnothrissa miodon* fish species locally known as *kapenta*. Artisanal fishing is non-mechanised and low cost, with fishers targeting multiple fish species inshore. Fishers also generally use gillnets in small dug-out canoes. Further, the contribution of artisanal fishing does not get reflected in the gross domestic product (GDP) of either Zambia or Zimbabwe. Nevertheless, it importantly caters to local employment and livelihood needs as well as nutrition and food security. Engaging in this trade makes sense, considering the above-outlined environmental, socioeconomic and cultural conditions in which the inhabitants of the Valley live – conditions exacerbated by employment losses experienced by many due to SAPs. In fact, the SAPs effectively drove such populations from formal employment into the informal sector, part of which consists in artisanal fishery (see, e.g., Jul-Larsen et al., 2003).

The artisanal fishing industry is also relatively easy to enter. Because of this, some women and men have reportedly left their jobs to work in the sector on a full-time basis (Machena, 1990). All they need, according to Machena (1990: 8), is a hawker's licence. In Lake Kariba fishing communities, women are particularly known to actively participate in activities ranging from fish processing to marketing, along the lakeshore (Hachongela, 1997). Research reports confirm that "women often play a significant role in post-harvest processing and marketing" (Ndhlovu et al., 2017: 2209). This is despite encountering such challenges as traditional barriers that prevent them from taking part in fishing as well as inequalities in fishing permits allocation by relevant authorities (ibid.).

Women are also indirectly involved in the fishing industry at Kariba. For example, as far back as 1962, Cliggett et al. reported on field work they conducted in the Gwembe Valley, in the post-flooding period (following the construction of the dam). They found that the commercial Kariba fishery had, by then, already become the most important income source for thousands of people there. In relation to women, they observed that:

> Large fish camps played a major role in integrating women (who had begun to commercialize the sale of beer within their respective villages) into a market economy. Some took up residence in the fish camps to brew beer. Others came from surrounding villages to sell eggs, fowl and other produce.
>
> (Cliggett et al., 2007: 25–26)

The foregoing also sets the context in which this chapter seeks to grasp the influence of state power (in the framework of the global agenda for sustainable development) on the activities of grassroots actors in contiguous borderlands of proximate Southern African states. Specifically, it investigates the extent to which sustainable development goals (SDGs) filter down, by way of state policies, to the grassroots in the borderlands of the selected SADC member countries. Principally, SDG 14 is applicable to Lake Kariba, though it is an inland waterbody. SDG 14 addresses life below water, aiming to sustainably manage and protect marine and coastal ecosystems. SDG 14 Target 4 specifically aims to:

> By 2020, effectively regulate harvesting and end overfishing, illegal, unreported and unregulated fishing and destructive fishing practices and implement science-based management plans, in order to restore fish stocks in the shortest time feasible, at least to levels that can produce maximum sustainable yield as determined by their biological characteristics.

Further, SDG 14 Target 9 seeks to "Provide access for small-scale artisanal fishers to marine resources and markets". Thus, the same SDG (14) includes two targets which address two key issues that constitute the core of the inquiry in this chapter.

That is, whether power over the space within specified territorial boundaries rests with the state; and whether the state sets out the agenda for realising the goals of sustainable development in that territory. Then, this chapter examines the ways in which grassroots actors conduct socioeconomic activities in order to identify those which reflect the actors embracing, or not, state policies designed to ensure that sustainable development is achieved. Hence the question, do grassroots actors in contiguous border areas who exploit shared transboundary resources embrace state measures designed to regulate the use of those resources? To what extent do they adhere to the regulations? The chapter thus focuses on the water sector and grassroots actors engaged in fishing on transboundary waterbodies, to examine their socioeconomic activities. The activities are examined against policy measures aimed at promoting sustainable fishing, especially emanating from or on the Zambian side of the border, in view of the national fishing ban which the Government of Zambia imposes annually.

Understanding human activities around Lake Kariba: definition of key terms and approach

The fact that the fishing ban is a policy regulation imposed on the Zambian side only, directs attention to the Zambezi River and Lake Kariba, as a natural boundary and one where the international border between Zambia and Zimbabwe is located. The boundary also marks the space where the jurisdiction and power to impose policies made within either country's domestic political economy ends. In this case then, the *border* between Zambia and Zimbabwe would be defined as a line at the margins of their respective sovereign territories which separates them from one another. Dynamics of inclusion and exclusion sometimes play out at this site, as discussed later in this chapter. However, the interactions at Lake Kariba are generally inclusive, cooperative and integrative. Of course, processes that generate tension and conflict among grassroots actors as well as between grassroots actors and state agencies do exist. But because of its social and cultural history, the Kariba region is overall liminal; linking the cross-border communities of mostly Tonga origin. The peoples of this ethnic group on either side of the border were previously only separated by the Zambezi River, before the colonisers established Southern Rhodesia and Northern Rhodesia. Thus, transnational identity and intergroup dynamics define the boundary at Kariba. The local culture there effectively bridges the international boundary. And this is because the cross-border communities largely share the same history and culture – that is, language, ethnicity and place of belonging. So, the "borderlands" communities around Lake Kariba are in essence unified by a common identity which facilitates cross-border coexistence, somewhat erasing the boundary (cf., Brunet-Jailly, 2005; Grimson, 2012). As a *boundary*, the river and Lake at Kariba could thus be described as other lines between and within countries centred on *inter alia* ethnicity and culture. Moreover, besides being a natural border, the Zambezi River, which was dammed to form the Kariba water reservoir, might ordinarily be considered a *frontier*, particularly if unconquered or unexplored. So too might Lake Kariba. But then, the lake covers a previously politically determined international border, which is also naturally defined as such at some sections. This is beside the fact that the lake *is* navigable and navigated by various actors, including the artisanal fishers, tourists and tour operators, managers of the Kariba hydroelectricity complex, and other locals. Some of the several islands on the lake are actually also inhabited by fishers and other residents. But the spatial reach of the power and authority exercised over citizens within the two respective territories that share the Zambezi River and Lake Kariba ends at the line demarcating the territories. That is, the enforcement of policies and regulations which are drawn on either side of this line only respectively apply in the territory defined by and within the bounds of that line. Adherence to such policies follows suit. For its purposes, this chapter conceives of a *policy* as a statement of intent or course of action taken by an actor such as a government towards attaining specific goals. A working example is the annual fishing ban effected in Zambia. The Government of Zambia regulates the country's fisheries sector through the Ministry of Agriculture and Livestock. Within the ministry, the Fisheries Department operationally manages the fisheries sector. Under the Fisheries Act, 2012 [No.22 of 2011] the government imposes an annual "prohibited fishing period", the rationale of which is to give ample time for depleted natural resources (i.e., fish) in the country's lakes and rivers to be replenished. Effectively, the fishing ban is a closed season for fishing in which fish are allowed to spawn in order to avoid depletion. This policy regulation is imposed annually for three months, from 1 December to the last day of February the following year. Engaging in any fishing or possessing fish during the fishing ban period is an offence. As concerns the overall objective of this chapter, policy regulations such as the fishing ban as well as the extent to which artisanal fishers embrace those courses of action are examined to understand if they contribute towards sustainable development. If attained, this would translate into achieving the SDGs.

In order to realise this objective, a thorough review of international, regional and the policies, legislation and programs of the Governments of Zambia and Zimbabwe as well as NGOs that operate in both countries and represent/work with Kariba region communities is conducted. The activities of artisanal fishers are examined against these instruments, which are proxies for state influence. Reports on fishing and related activities concerning artisanal fishers from newspapers, NGOs, members' associations and so on also provide data sources for the analysis. The activities of the grassroots actors as thus reported are assessed against the instruments to determine adherence or non-compliance with policies designed to achieve sustainable development. The rest of the chapter is divided into four sections.

The first introduces the context of the discussion and issues addressed in this chapter. This is done by providing a brief overview of continental, regional and (Zambian and Zimbabwean) national fisheries legislation and policy frameworks. The second demonstrates the way in which fisheries regulations in Zambia and Zimbabwe affect artisanal fishers. This is accomplished by examining five issues that are gleaned from those regulations. The third starts by indicating that Zambia and Zimbabwe are yet to incorporate SDGs in their respective domestic fisheries policies. It then goes on to argue that special interests play significant roles in shaping fisheries legislations in both countries. The last section concludes and makes the recommendation for proximate states that share transboundary waterbodies to consider developing localised frameworks of governing local resources.

Continental, regional and national fisheries legislation and policy frameworks: a brief overview

Fisheries and other legislation, policies and regulations that touch upon Zambia and Zimbabwe's respective fisheries sectors were formulated and imposed before many relevant global, continental and regional instruments. This is true vis-à-vis *inter alia* the SDGs, the African Union (AU)/New Partnership for Africa's Development's (NEPAD) *Policy Framework and Reform Strategy for Fisheries and Aquaculture in Africa*, the 2000 Revised Protocol on Shared Watercourses in the SADC, and the 2001 SADC Protocol on Fisheries.

Besides the SDGs, some global instruments have specific interest in sub-themes of the fisheries sector, such as combating illegal activities to protect fish resources. But such instruments, too, ultimately seek to contribute to sustaining fish resources. An example is the International Plan of Action to prevent, deter and eliminate Illegal Unregulated Unreported fishing (IPOAIUU). The international scene also includes some international NGOs and agencies that also operate in Zambia and Zimbabwe, and address aspects of the fisheries sector. These too aim to sustain and conserve fish resources. For example, the World Fish Centre (WFC) established in 1977. It is concerned with conserving and sustaining fish resources out of its commitment to reducing poverty and hunger by improving fisheries and aquaculture (World Fish Center, 2018). Similarly, the World Wide Fund for Nature (WWF) works in Zambia and Zimbabwe to *inter alia* ensure that the most important fisheries ecosystems are productive and resilient and improve livelihoods and biodiversity; freshwater ecosystems and flow regimes provide water for people and nature; and that sustainable food systems conserve nature and maintain food security (World Wide Fund For Nature, 2018). Such organisations mostly engage in clarifying prohibitions and regulatory instructions in order to maximise the sustainable use of fish resources.

Continentally, instruments such as the Policy Framework and Reform Strategy for Fisheries and Aquaculture in Africa aim to facilitate coherent policy development in AU member states for sustainably managing fisheries and aquaculture resources (African Union, 2014). Regionally, the Revised Protocol on Shared Watercourses in the SADC aims for integrated water resource management between member states. This approach to water and water resources management is considered easier to achieve when conducted outside the limits of an individual SADC member state's borders (SADC, 2011). Besides the shared watercourses protocol, the SADC Protocol on Fisheries specifically seeks sustainable and responsible use of aquatic resources and ecosystems of interest to member states. Its aims include *inter alia* promoting food security and human health; safeguarding the livelihoods of fishing communities; generating economic opportunities for nationals in the region; ensuring that future generations benefit from these renewable resources; and alleviating and eradicating poverty (SADC, 2001). Several other SADC instruments too, such as the Regional Agricultural Policy (RAP) and the Regional Indicative Strategic Development Plan (RISDP), provide strategies, guidelines and regulations to foster cooperation in and the performance of fisheries in member states. Their ultimate aim is to improve the welfare of member states' citizens and overall development of the regional economic community, member states and Africa.

It is also noteworthy (vis-à-vis this chapter's objective) that in their quest for stated goals, some of the highlighted instruments emphasise policies and regulations. But they also give some attention to grassroots actors. For example, the Policy Framework and Reform Strategy for Fisheries and Aquaculture in Africa includes a principle to develop artisanal fisheries – to contribute to the alleviation of poverty. The framework actually acknowledges that artisanal marine fisheries are the highest producers among the fisheries sub-sectors, accounting for 0.42 per cent of the total US$24 billion value added from Africa's fisheries and aquaculture sector (African Union, 2014: 6). At the national level, Zambia is presently drafting a fisheries policy, based on the Fisheries Act, 2012 [No.22 of 2011] cited earlier. In the meantime, other national policies address issues in fisheries. For example, the

Second National Agricultural Policy (2016) includes fisheries as a sub-sector and has put in place policy objectives and measures based on the Fisheries Act to increase production and productivity. The Zambezi River Authority Act cited earlier also makes provisions for the use and management of the Zambezi River in Zambia and Zimbabwe.

For Zimbabwe, the principle legislation that governs the control, management and development of its fisheries is the National Parks Act, 1975 [Parks and Wildlife Act–Chapter 20:14 of 1990 and 1996, as amended]. The legislation includes fisheries management regulations too. This is noteworthy for two reasons. Firstly, because part of the Act deals with the conservation of fish. Secondly, because, as explained later, the regulations seem to rather focus on placing restrictions on artisanal fishers' activities. According to the Act, such restrictions are designed to regulate fishing effort and include limiting the number of licenses for fishing, and controlling and restricting fishing areas, fishing gear, fishing net mesh size and the number of fishing nets each fisher can use (see, Part XIV, Parks and Wildlife Act, 1975). The custodian of the Act is the Ministry of Environment and Tourism. This means the final authority over Zimbabwe's fisheries resources inheres in the Minister, through the Director of the Department of National Parks and Wildlife Management (DNPWM). The Director is thus empowered to control, regulate, prohibit or restrict fishing in the controlled waters of Zimbabwe.

Like Zambia, no fisheries policy exists in Zimbabwe. Instead, Zimbabwe's Agricultural policy and the Livestock production policy make some provisions for water and aquaculture development. Further, other pieces of legislation and policies such as the Inland Water Shipment Act, include frameworks that affect fisheries or management of the Kariba reservoir and the Zambezi River. Also, the DNPWM has an overall strategy of using fisheries sustainably to ensure that biological diversity is not lost (FAO (Food and Agriculture Organisation of the United Nations), 2003). The department is also engaged in efforts to increase fish production to strengthen rural economies, create employment and promote food security. Besides this, the Parks and Wildlife Act and subsequent Statutory Instruments outline some legal provisions of the country's fish licensing system (FAO (Fisheries and Aquaculture Department), 2016). And because all fisheries belong to the state, the Act invests all the responsibility for their management in the state.

State power meets the grassroots: fisheries regulations and artisanal fishing activities, the interplay

The way in which artisanal fishers on either side of the boundary at Lake Kariba are affected by and respond to fishing regulations can be considered through five issues. These include fishing period, fishing grounds, fishing equipment and methods, market for artisanal fishers' produce and what I call *stray fishing*. The fishing ban is enshrined in Zambia's Fisheries Act. The Act, as amended by statutory instruments, provides detailed listings of specific fishing areas in Zambia, including rivers, lakes, wetlands and other water-bodies where fishing is prohibited, the extent of the prohibitions and the periods in which those prohibitions are applicable. Indications are that "none" of the prohibitions apply to the "Kariba dam/lake" (Zambia Legal Information Institute, 2018). Instead, it is the section of the Lower Zambezi River from the Dam wall to the Zambezi/Luangwa Confluence which is subject to the 1 December to 28 February period of (fishing) prohibition.

This means that the Zambian fishing ban does not apply to the Kariba water reservoir per se, but to the body of water below the dam up to where the Zambezi River converges with the Luangwa River – at the tripartite border between Zambia, Zimbabwe and Mozambique. Fishing on the lake is thus carried out throughout the year. Thus, the fishing regulations that respectively apply in Zambia and Zimbabwe do not apply to the trans-boundary waterbody at Kariba. So, while Zambia experiences constant mobility of artisanal fishers along the lakeshore of the Kariba and "from the shore to a great number of unpopulated islands [on the lake] from where the fishermen [...] freely seek what they consider to be some of the best fishing grounds" (Jul-Larsen, 2003: 235), the same cannot be said about Zimbabwe, as detailed shortly below. The bottom line here though is that, on the Zimbabwean side of the boundary, state law principally bans artisanal fishers from fishing. And where they are permitted to fish, the law limits the kind of gear they can use and the quantities of fish they can draw. Those permitted to fish do so under a system in which their local governments are handed down the designated number of permits for a given year.

This situation also raises concerns that, if the Kariba remains open to Zambian artisanal fishers (who happen to be generally free to fish anywhere along the lake shoreline) during the Zambian fishing ban, pressure on the lake and its fish resources might increase. This is because, fishers from other fishing areas subjected to the ban might want to come and exploit the only grounds in the country that are open to fishing during the period. This might lead to an increase in the number of fishers on the lake and strain the fish resources. So, the effect of state policy or the consequence of embracing it, particularly on the Zambian side, translates into fishers finding an alternative space to fish during the fishing ban, in the Kariba, which remains open. This effectively increases pressure on and threatens fish stocks, with consequences for the management of this resource. Some studies, though, have shown that Lake Kariba faces no threat of depletion of its fish resources from overexploitation (see, e.g., Jul-Larsen, 2003; Kolding et al., 2016). Also, although the regulation says anyone caught with fish in Zambia during the fishing ban is guilty of an offence, it could be argued that the Zambian state exhibits some laxity on this point. This is because the state seems inconsistent by applying the ban to some of the country's waterbodies while leaving out Lake Kariba.

Variations exist between the Zambian and the Zimbabwean sides of the boundary at Lake Kariba pertaining to the grounds where artisanal fishers are permitted to fish. As hinted above, the local and indigenous population on the Zambian side is free to fish anywhere along the Kariba shoreline (Jul-Larsen et al., 2003). This liberty dates back and owes to the fact that from the time they were displaced by the flood, efforts were made by the state and local authorities in Zambia to promote fishing among the displaced people as an alternative to farming, which used to be their main economic activity and primary means of subsistence. Besides the flood, rising pressure on and demand for land where these populations had been resettled negatively affected agriculture. Add to this, the harsh physical environment and adverse weather patterns discussed earlier, which negatively affected harvests. Thus, locals were encouraged to go into fishing, to release the pressure and provide them with alternative livelihoods and means of survival.

The situation is different on the Zimbabwean side where part of the lake's shoreline is reserved for tourism, with the lake itself being designated a recreational park. Most of the dry land adjacent to the lake too consists of national parks and safari areas. Thus, unlike the entire shoreline open to artisanal fishing in Zambia, only about 60 per cent on the Zimbabwean side is open to this economic activity. And this too dates back to the time Zimbabwe was under colonial rule, as elaborated later. The resident population, who constituted the majority of artisanal fishers, were allotted "Native Areas" along the shoreline where they could fish inshore but were prohibited from permanently settling there or engaging in shoreline agriculture (Kolding et al., 2003: 72). Because most of the land adjacent to Lake Kariba was earmarked for national parks, some of the waters that were adjacent to these tourist spots were, accordingly, not open to artisanal fishing. In fact, artisanal fishing was essentially not allowed, as the indigenous people were encouraged to turn professional. To date the DNPWD in Zimbabwe controls inshore fishing on Lake Kariba (and all other waterbodies in the country under its control). The Department closes areas and limits access to fishing and restricts the type of fishing gear and methods used (Kolding et al., 2003) by artisanal fishers. Further, as Kolding et al. (2003) indicate, DNPWD runs a permit system in which it informs the local authorities on the Lake (e.g., Nyaminyami and Binga District Councils) of the limits on the number of fishing permits the councils may issue to fishing cooperatives or individual fishers in a particular year. The history and rationale behind this system, which has "never been reviewed" (Kolding et al., 2003: 73), goes back to Zimbabwe's pre-independence period in which fishing interests were racially segmented and fishing grounds and Lake Kariba waters were largely allotted to the conservation of stocks to meet demands for recreation in the tourist industry. Simultaneously, the total area made available to inhabitants of the Gwembe Valley for inshore fishing was reduced.

The foregoing goes to emphasise that fishing regulations on the Zimbabwean side of Lake Kariba promote sport-fishing, as a tourist undertaking. The lake is a recreational park and any artisanal fishers found fishing there could be charged with targeting threatened fish stocks and species that are meant for this pastime (Jul-Larsen et al., 2003). Even present day independent

Zimbabwe's DNPWA establishes exclusive fishing zones in which artisanal fishers have to decide among themselves on who and how to use the quotas allotted to them at Lake Kariba. And while "the Zimbabwean side has been strictly managed and controlled in terms of licensed fishers, restricted fishing grounds, and minimum mesh-size regulations, the Zambian side has virtually been an open-access fishery with no enforcement of regulations since its independence in 1964" (Kolding et al., 2016: 646).

Fishing equipment seems to generate conflict between Zambian and Zimbabwean fishers at Lake Kariba. The conflict apparently emanates from differences in applicable regulations on the authorised size of fishing nets for fishers on either side of the border. The smallest legally permissible fishing net mesh size for fishers on either side of the border at Lake Kariba is the same: 8 mm. It is worth noting, however, that this size is prescribed for *kapenta*, which, as explained earlier, is exploited offshore and, mostly, by predominately white-owned, capital intensive and highly mechanised commercial ventures that have turned it into a million-dollar industry (Overå, 2003; FAO, 2003; Kolding et al., 2003, 2016; Zambia Legal Information Institute, 2018). The legal size prescribed for the gillnet, which is predominately used inshore by artisanal fishers on either side of the border, is 76 mm in Zambia and 102 mm in Zimbabwe (FAO, 2007; Zambia Legal Information Institute, 2018). The smaller size permitted to Zambian fishers implies that they can draw more fish than their Zimbabwean counterparts. But also, Zambian fishers pose a greater threat to fish stocks in the lake, as their nets draw juveniles as well as adults.

This minimises chances for the fish to grow to maturity and spawning age and, therefore, threatens the sustainable growth of fish populations. Further, and bordering on illegality and the extreme, some fishers reportedly use illegal and unconventional fishing methods. The fishers allegedly use mosquito nets, which are smaller in mesh size than even the 76 mm permitted in Zambia and complained against by Zimbabwean fishers. Reports of artisanal fishers using mosquito nets say they use them for lack of raw materials to make legally acceptable nets (Siamonga, 2014). However, such nets violate the legal provisions of the Fisheries Act, 2012 (Zambia) and the Parks and Wildlife Act, 1975 (Zimbabwe). Both Acts limit mesh sizes to prevent juveniles and small fish species from being harvested and collapsing their stock. There are also reports of fishers using chemicals as a fishing method (Kandawire and Samboko, 2017). The chemicals pose a threat to not only the fish they kill but other micro and macro marine organisms too. They also alter and pose a threat to the ecosystem. Clearly, artisanal fishers are found wanting on the count of employing prohibited and unconventional fishing equipment and methods. Non-adherence to policy is very evident in these cases.

Further, the regulations on the Zambian side do not impose any restrictions on the number of fishing nets an artisanal fisher can own. Not so on the Zimbabwean side, where each fisher is restricted to five nets only. This implies that there are more nets on the Zambian side than the Zimbabwean and, therefore, the density increases the catch in Zambia. This, combined with net size and lack of entry regulations for aspiring fishers into the fishing sector, implies a greater threat to sustainable use of fish resources on that side of the lake. However, some studies show that no threat of overexploitation of fish exists on Lake Kariba (Jul-Larsen, 2003; Overå, 2003).

Still more, the Zambian side seems to have more fishers than the Zimbabwean side. The number of fishers operating on Lake Kariba is, again, determined by regulations on access to the lake's resources in the respective countries. As hinted, in Zambia, the fisheries sector is generally open to anyone interested in fishing. A slight local variation to this exists at Kariba though. This can be traced back to the 1990s when both Zambia and Zimbabwe introduced co-management in the sector, aimed at decentralising management to institutions at the local level (see also, e.g., Malasha, 2003; Overå, 2003; Ndhlovu et al., 2017). This led to the establishment of Zonal Management Committees (ZMCs) and Sub Area Fishermen's Associations (SAFAs) on the Zambian and Zimbabwean sides of the lake, respectively. The ZMCs, which also established Integrated Village Management Committees (IVMCs) below them, address fisheries and land planning issues under the chairpersonship of traditional chiefs. The IVMCs on their part exercise control over access to the fisheries and ensure fisheries management regulations are observed and complied with. They include a chair elected from among the fishers, a village head, three ordinary members (elected), a Department of Fisheries-appointed village scout and a fisheries assistant.

Across the boundary, SAFAs, unlike ZMCs, were mandated to help impose fisheries regulations and the collection of data for monitoring purposes. Though each SAFA comprised four to six adjacent villages, no traditional leaders, artisanal fishers or any other Valley inhabitants who use the lake's resources are included in these local institutions. Instead, the three individuals who are nominated to each SAFA have an exclusive zone to themselves to manage, help with law enforcement and collect fishing data. Zimbabwean artisanal fishers complain that their Zambian counterparts outcompete them by flooding and illegally selling cheaper fish in Zimbabwean markets. Reports of Zambian fishers crossing the Kariba into Zimbabwean towns such as Binga and informally or illegally selling their catch there are common (see, e.g., Chiutsi, 2015). Because of the less restrictive artisanal fishing regulations and relatively easier environment in which Zambian fishers operate in their own country, they seem to have a clear advantage over Zimbabwean fishers, as they have more and cheaper fish to offload on the market. But also, the physical infrastructure and fish marketing and distribution networks on the Zambian side are not comparable to those across the border, making the Zimbabwean side attractive.

The Zambian side of the Gwembe Valley has not improved in both physical infrastructure and fish marketing and distribution networks that link and could facilitate transportation of fishers' produce to major Zambian cities north of the Valley such as Monze, Mazabuka and Lusaka. And this does not significantly differ from the general observation which Cligget et al. made of the area in 2007:

> Roads deteriorated and in some cases vanished. Agricultural inputs were delivered late if at all, and many areas found buyers unwilling to collect crops because this was uneconomical due to the bad roads. Prices for food crops were uncertain, especially when those trying to sell grain had to compete with grain imported from the south or as relief food from overseas.
>
> (Cliggett et al., 2007: 23)

Zimbabwe, however, has better local economies and marketing and distribution infrastructure centred around such towns as Binga and Mlibizi. Further, that side of the border is easily accessed by customers from major cities such as Bulawayo. Although Zimbabwean fishers complain about the competition for their markets and facilities, it is actually acceptable practice for Zambians to cross into Zimbabwe twice a week to sell their fish at flea/open markets in Binga, for instance. The situation is, nonetheless, a source of conflict among artisanal fishers in the Kariba borderlands.

Besides practices that cause cross-border conflict between artisanal fishers in the Kariba borderlands, artisanal fishers also have conflicts with state authorities or agents. The artisanal fishers also often stray into foreign waters, across the international border on Lake Kariba, only to get arrested for illegal crossing and fishing. *Stray fishing* is a violation of the international boundary between Zambia and Zimbabwe. This is because the fishers wander and find themselves fishing in foreign waters without any passports or border passes. Stray fishing is also a graver violation for fishers caught on the Zimbabwean side of the lake because artisanal fishing is fundamentally prohibited on that side and considered a threat to sport-fishing, which is largely the use for which Zimbabwean regulations reserve the lake (see above).

So, how do the riparian states handle issues of stray fishing in the borderlands of Lake Kariba? Stray fishers on Lake Kariba are usually arrested both ways. This is because fishers from both sides of the border stray into the other country's territory. Newspaper reports suggest that such incidents generally end up being mutually resolved by the state authorities stationed around the lake, with counterparts across the border, and the stray fishers are usually released. Importantly, this all implies the existence of some form of understanding and acceptance of the phenomenon between the authorities on either side of the border as well as, even, the fishers and Valley inhabitants in general. Else, extreme cases would be resolved diplomatically in the context of existing regional and international legislation such as the SADC Protocol on Shared Watercourses (see, e.g., Article 7) or bilaterally under Joint Commissions of Cooperation (JCC). However, the way in which cases of stray fishing are handled between Zambia and Zimbabwe at Lake Kariba has so far not exceptionally escalated to even require the intervention of authorities in their respective capitals. This is also indicative of the existing local social and cultural affinities between the cross-border communities in the Gwembe Valley. Further, it suggests that the border between the two countries at this location effectively separates the same people. Or that the border is effectively non-existent insofar as concerns the inhabitants of the Valley. The literature on border studies would, therefore, deem this border to be essentially bridged (Wilson and Donnan, 1998; Donnan and Wilson, 1999; Wilson and Donnan, 2005). And this, because of sociocultural similarities between the peoples which the border separates.

The absent global-cum-regional goals for sustainable development in Zambia and Zimbabwe: an exposé from the Kariba borderlands

Some of the highlighted activities and methods of fishing employed by the artisanal fishers in Lake Kariba threaten the sustainability of the lake's fish resources. However, this chapter has in an unspoken but salient way, shown that the measures designed to regulate those activities are at present actually not directly determined by SDGs. Though some of them somewhat resonate with the new global goals, they are rather informed by legislation drawn long before the SDGs. Yes, the basic fisheries legislation in Zambia and Zimbabwe has not fundamentally changed from that drafted during the colonial era. And it does not speak to the observable realities and conditions in present day Gwembe Valley. It is outdated, suited for a different epoch and regime, and out of touch with the realities and cultures of the people whose lives and livelihoods the global, continental and regional instruments, as well as the work of some international NGOs cited in this chapter, seek to promote. Clearly, that legislation needs updating if it is to be aligned to the SDGs and the needs of artisanal fishers.

Furthermore, within Zambia and Zimbabwe, national and special interests seem to determine the kinds of regulation drawn for the management of the respective countries' fisheries sectors. The [p]reservation of Lake Kariba as a recreational park on one side of the border and concerted efforts to push locals to take up fishing (Jul-Larsen, 2003) as an occupation and means of livelihood on the other side of the same border says a lot about the interests that shape national policymakers' priorities. During colonial rule on the Zambian side, the authorities were faced with pressure on upland as resettled populations from the flooded Gwembe Valley grew and not only posed a threat to local social and cultural stability and food security, but national food security and economic production too. No such pressure or threats seemed to exist on the other side of the border in Zimbabwe. This led to a divergence in regulations affecting artisanal fishers. Looking back at the respective historical development of fisheries legislation in Zambia and Zimbabwe since the colonial period, Malasha (2003: 254) is of the same view in positing that although the legislation seemed to be based on scientific principles, its implementation was influenced by the political and economic interests of the state.

Specifically, the scientific and ideological justifications for the regulations imposed in these respective territories disguise the political and economic interests of specific groups that pushed for their implementation, as the laws were essentially designed to satisfy the interests of the groups that dominated these countries. According to Malasha (2003), the evolution of fisheries regulation in colonial Zambia was tied to copper production. One time during the colonial period, the Zambian Copperbelt[2] had experienced massive shortages of beef. The economic groups that controlled the colony's mining industry then had argued that the situation threatened instability among mineworkers and would affect copper production if allowed to continue. In response, the colonial government eased restrictions on fishing regulations throughout Zambia to allow fish traders distribute fish from across the country's waterbodies to the Copperbelt Province. Additionally, the authorities realised they could make use of the revenues collected from fish licensing to *inter alia* finance administrative structures in rural areas as well as pay local chiefs' salaries.

Across the Kariba in Zimbabwe, Malasha (2003: 261) traces the special interests that marginalised Africans from fishing – holding the Africans' fishing methods to be destructive and unsportsmanlike – to fisheries legislation made back in 1881. The special interests would eventually become a sport-fishing lobby, which not only imported exotic fish species but also pressured government to finance such imports as well as legislate protection for the lobby's sport. Legislation was also made, at the aegis of this lobby, to ban fishing for certain periods of time to allow fish to increase. Thus, elite clubs and associations drove the development of fisheries regulation in Zimbabwe. Fisheries legislation in Zambia and Zimbabwe has not undergone any fundamental change from what colonisers crafted before the countries respectively attained political independence. This is despite the various amendments to and the numerous subsequent statutory instruments that have since been drawn from the legislation. Else, a level of standardisation in the application and results of global-cum-regional goals which are translated into national policy and local regulations which govern artisanal fishing on one hand, and grassroots activities which are consistent with such policy and regulations on the other hand, would have been evident in Zambia and Zimbabwe today, on the Kariba transboundary waterbody. The application of such measures and regulations should indeed yield comparatively similar results on both sides of Lake Kariba. This is an expectation which is not far-fetched. Policy harmonisation is a common concept in regional integration parlance and practice.

To borrow an example of its application from the labour sector, the SADC Regional Labour Migration Action Plan 2013–2015 prescribes *inter alia* that member states should develop a comprehensive strategy to address mixed and irregular migration in the SADC region (SADC Secretariat, 2015). The Regional Action Plan has six key pillars which are elaborated into specific actions, time frames, indicators, implementation budget and so on that member states individually (i.e., domestically) and jointly (i.e., the collective of all 16, working together) implement by developing respective National Action Plans that are consistent with the framework of the Regional Action Plan. The work and contribution of each member state conducted in this way culminates in a common regional approach (of contributions by all) that responds to and addresses irregular and mixed migration. In this way, respective SADC states have cooperated under the SADC Secretariat to draw and implement national policies, programmes and actions drawn from regulations based on the Regional Action Plan. The efforts of member states and the harmonisation of their respective national policies with the six key outlined pillars of the Regional Action Plan, and the progress they make in implementing these measures are even trackable through evaluation and monitoring mechanisms at the SADC Secretariat's Employment and Labour Sector (ELS).

Grassroots realities embracing or countering state power in borderlands: concluding notes

This chapter has attempted to gain insight into the influence of state power in the context of the global agenda for sustainable development on the activities of artisanal fishers who inhabit borderlands on Lake Kariba, the transboundary waterbody shared by Zambia and Zimbabwe. The chapter has investigated the degree to which state fisheries legislation and regulations filter down to and affect artisanal fishers in the borderlands of the two countries, which are also member states of the SADC. The assumption was made that, if power over the space within specified territorial boundaries rests with the state, and if the state set out the agenda for realising the goals of sustainable development in that territory, then, the ways in which artisanal fishers conducted their socioeconomic activities was worth investigating in order to identify those activities which show the actors (not) embracing state policies designed to ensure that sustainable development was achieved. The fishers' activities were examined against policy measures aimed at promoting sustainable fishing that emanate from or on the Zambian side of the border at Lake Kariba, considering the national fishing ban which the Government of Zambia imposes annually. The legislative and policy framework of global, African, regional and (Zambian and Zimbabwean) national fisheries legislation and policies provided the context in which the chapter discussed the regulations and artisanal fishers' activities at the common Zambian–Zimbabwean border.

The chapter has thus demonstrated that fisheries regulations in Zambia and Zimbabwe affect artisanal fishers. This happens through five key issues gleaned from those regulations, comprising: the period in which fishers are permitted or prohibited from fishing, the grounds or areas where they are permitted to conduct their fishing, the equipment and methods they are allowed by law to employ for fishing, issues regarding markets for their catch and, lastly, the phenomenon in which the fishers occasionally stray and are caught (or arrested for) fishing in foreign waters. Further, the chapter has highlighted the fact that Zambia and Zimbabwe are yet to incorporate the SDGs in their respective domestic fisheries legislation. Currently, the fisheries legislation and regulations in these countries are fundamentally based on legislation passed when they were both under colonial rule. The legislation itself was, in turn, significantly shaped by special interests during that dispensation. The legislation also essentially fails to consider the interests of artisanal fishers, who really should gain the most from the resources of Lake Kariba, especially considering their historical experience and that of other inhabitants of this area, of forced and disruptive displacement induced by the construction of the Kariba dam. Before the forced displacement and relocation, these people self-sufficiently engaged in subsistence agriculture and lived off the Zambezi River and its tributaries.

Moreover, way before that and prior to being separated by the colonial Rhodesian government through the construction of an international political line demarcating Zambia and Zimbabwe, the inhabitants predominately consisted of one ethnic group whose communities cut across the Zambezi River, the only existing and natural boundary then. Any attempts to improve the lives of artisanal fishers and the inhabitants of these borderlands in general should thus take the history and these realities into consideration. This is important for whatever solutions are found for the sustainable use and management of Lake Kariba to be meaningful. In view of this and the discussion in this chapter, the general proposition and recommendation is made that proximate states that share transboundary waterbodies in contiguous border areas need localised governance frameworks to manage the resource. In practice, this could involve the deliberate harmonisation of national and local legislation, policies and regulations in the context of regional and global instruments. But if bilaterally pursued, this should be conducted under JCCs, which are a standard feature of inter-state collaboration.

Further, locals and especially traditional leaders, must form part of and be included in such institutions and processes. Such participation should, of course, constitute both male and female local leaders. Local participation is important in order to ensure that the original inhabitants – in this case, of the Gwembe Valley – are not marginalised (as they allege has always been the case). But also, as Muriritirwa (1995) has observed, marginalising traditional leaders could result in the institutions so established at local level being weak and unable to *inter alia* mobilise and facilitate participation in managing local resources. It is thus imperative to appreciate local grassroots socioeconomic activities and cultural dynamics. Also, local livelihoods ought to be promoted as alternatives in order to curb unregulated or ill exploitation of common and transboundary resources. For example, the use of chemicals for fishing.

I conclude by recollecting the observation that while the state has power to impose legislation in its territory, the design of such legislation is influenced by and represents certain special interests that operate within the territory of that state. As if fully cognisant of this, grassroots actors such as artisanal fishers sometimes go against such legislation in the practice and conduct of their vocation. Not only so, the culture in which the actors are embedded seems to generally understand such practices and to also ride within the boundaries of the culture itself and the limits of the practices; as if to counter political machinations that design boundaries and regulations which are seemingly aimed at constraining the grassroots actors of those cultures. The common cultural understanding among the grassroots actors also seems to extend to the local agents of the state in the borderlands, as exhibited in the way in which those agents in the Gwembe Valley reportedly resolve problems that arose from the violation of state regulations. This is to be expected, as those agents operate within the same social, cultural and political boundaries as the grassroots actors.

Notes

1 Lord Llewellin (1956) announced to a joint meeting of the Royal African Society and the Royal Empire Society on 5 July 1956 that the first current from the then new hydroelectric scheme on the Zambezi River – to satisfy the demand for electricity on the Copperbelt in Zambia (then Northern Rhodesia) – would be available in 1960. The scheme dated back to 1951, when the Inter-Territorial Hydro-Electric Commission "recommended the development of a dam at Kariba and hydro-electric power station". A brief history of the ZRA is available on their website: www.zaraho.org.zm/history.html (accessed 14 April 2014).
2 Since the colonial period, copper remains the mainstay of Zambia's economy. Copper mining is concentrated in a province north of Zambia that borders the Democratic Republic of Congo (DRC) called the Copperbelt Province. Labour from the rest of Zambia and abroad has historically migrated to the region and thus contributed to population growth and the urbanisation of the province.

References

African Union. 2014. Policy Framework and Reform Strategy for Fisheries and Aquaculture in Africa. African Union. Inter African Bureau for Animal Resources. https://au.int/sites/default/files/documents/30266-doc-au-ibar_-_fisheries_policy_ framework_and_reform_strategy.pdf.

Brunet-Jailly, Emmanuel. 2005. Theorizing Borders: An Interdisciplinary Perspective. *Geopolitics* 10(4): 633–649.

Chiutsi, Ngonidzashe. 2015. Binga Fish Farmers Cry Foul over High Permit Fees. *The Sunday News*, 17 May. www.sundaynews.co.zw/binga-fish-farmers-cry-foul-o ver-high-permit-fees/.

Clark, Sam, Colson, Elizabeth, Lee, James and Scudder, Thayer. 1995. Ten Thousand Tonga: A Longitudinal Anthropological Study from Southern Zambia, 1956–1991. *Population Studies* 49(1): 91–109.

Cliggett, Lisa. 2001. Carrying Capacity's New Guise: Folk Models for Public Debate and Longitudinal Study of Environmental Change. *Africa Today* 48(*1*): 3–19.

Cliggett, Lisa. 2002. Male Wealth and Claims to Motherhood: Gendered Resource Access and Intergenerational Relations in the Gwembe Valley, Zambia. In G. Clark (Ed.), *Gender at Work in Economic Life*. Walnut Creek, CA: AltaMira Press, pp. 207–223.

Cliggett, Lisa, Colson, Elizabeth, Hay, Rodrick, Scudder, Thayer and Unruh, Jon. 2007. Chronic Uncertainty and Momentary Opportunity: A Half Century of Adaptation among Zambia's Gwembe Tonga. *Source: Human Ecology 35(1)*: 19–31.

Colson, E. 1960. *Kariba Studies: The Social Organisation of the Gwembe Tonga.* Manchester: Manchester University Press.

Donnan, H., and Wilson, T. M. 1999. *Borders. Frontiers of Identity, Nation and the State.* Oxford: Berg.

FAO (Fisheries and Aquaculture Department). 2016. Fishery and Aquaculture Country Profiles. Kenya (2016). Country Profile Fact Sheets. FAO Fisheries and Aquaculture Department [Online]. www.fao.org/fishery/facp/ZWE/en.

FAO (Food and Agriculture Organization of the United Nations). 2003. Information on Fisheries Management in the Republic of Zimbabwe. Zimbabwe. www.fao.org/fi/oldsite/FCP/en/ZWE/body.htm.

FAO (Food and Agriculture Organization of the United Nations). 2007. FAO Fishery Country Profile – The Republic of Zimbabwe. Rome. www.fao.org/tempref/FI/DOCUMENT/fcp/en/FI_CP_ZW.pdf.

Grimson, Alejandro. 2012. Nations, Nationalism and "Borderization" in the Southern Cone. In Thomas M. Wilson and Hastings Donnan (Eds), *A Companion to Border Studies*. Chichester, UK: Wiley Blackwell, pp. 194–213.

Groves, Zoë. 2013. Transnational Networks and Regional Solidarity: The Case of the Central African Federation, 1953–1963. *African Studies 72(2)*: 155–175.

Hachongela, P. 1997. *A Gender Analysis of Participation m Planning for Village Regrouping on Kariba Shoreline (Zambia).* CASS Occasional Paper – NRM Series; CPN 84/1997. Harare: Centre for Applied Social Sciences, University of Zimbabwe.

Jul-Larsen, Eyolf. 2003. Analysis of Effort Dynamics in the Zambian Inshore Fisheries of Lake Kariba. In Eyolf Jul-Larsen, Jeppe Kolding, Ragnhild Overå, Jesper Raakjær Nielsen, and Paul A.M. van Zwieten (Eds), *Management, Co-Management or No-Management? Major Dilemmas in Southern African Freshwater Fisheries. Part 2: Case Studies.* FAO Fisher. Rome: United Nations Food and Agriculture Organisation, pp. 233–252.

Jul-Larsen, Eyolf, Kolding, Jeppe, Overå, Ragnhild, Raakjær Nielsen, Jesper and van Zwieten, Paul A.M. (Eds). 2003. *Management, Co-Management or No Management? Major Dilemmas in Southern African Freshwater Fisheries. Part 2: Case Studies.* FAO Fisher. Rome: Food and Agriculture Organization of the United Nations.

Kandawire, Laiton, and Samboko, Oliver. 2017. What's Chocking the Kariba? *Patsaka Nyaminyami Community Radio*, 23 July. http://patsakacommunityradio.org/2017/07/23/whats-choking-lake-kariba/.

Kolding, J., Musando, B., and Songore, N. 2003. Inshore Fisheries and Fish Population Changes in Lake Kariba. In Eyolf Jul-Larsen, Jeppe Kolding, Ragnhild Overå, Jesper Raakjær Nielsen, and Paul A.M. van Zwieten (Eds), *Management, Co-Management or No Management?: Major Dilemmas in Southern African Freshwater Fisheries. Part 2: Case Studies.* FAO Fisher. Rome: Food and Agriculture Organization of the United Nations, pp. 67–99.

Kolding, Jeppe, Jacobsen, Nis S., Andersen, Ken H., van Zwieten, Paul A. M. and Giacomini, Henrique. 2016. Maximizing Fisheries Yields While Maintaining Community Structure. *Canadian Journal of Fisheries and Aquatic Sciences 73(4)*: 644–655.

Llewellin, Lord. 1956. Some Facts about the Federation of Rhodesia and Nyasaland. *African Affairs 55(221)*: 266–272.

Machena, C. 1990. *Lake Kariba Fisheries Research Institute Annual Report Project, Report Number 67*. Harare: Department of National Parks and Wildlife Management.

Malasha, Isaac. 2003. The Emergence of Colonial and Post-Colonial Fisheries Regulations: The Case of Zambia and Zimbabwe. In Eyolf Jul-Larsen, Jeppe Kolding, Ragnhild Overå, Jesper Raakjær Nielsen, and Paul A.M. van Zwieten (Eds), *Management, Co-Management or No-Management? Major Dilemmas in Southern African Freshwater Fisheries. Part 2: Case Studies*. FAO Fisher. Rome: Food and Agriculture Organization of the United Nations, pp. 253–266.

Ministry of Agriculture, and Ministry of Fisheries and Livestock. 2016. *Second National Agricultural Policy*. Republic of Zambia. http://cbz.org.zm/public/downloa ds/SECOND-NATIONAL-AGRICULTURAL-POLICY-2016.pdf.

Ministry of Environment and Tourism. 1975. *Parks and Wildlife Act, 1975 [Parks and Wildlife Act–Chapter 20:14 of 1990 and 1996, as Amended]*. Acts. Vol. *14*. Government of Zimbabwe. www.unodc.org/res/cld/document/parks-and-wild-life-act_ html/Parks_And_Wild_Life_Act.pdf.

Muriritirwa, W. 1995. *An Institutional Analysis of Co-Management Arrangements for Managing Gill Net Fisheries at Lake Kariba*. Harare: Centre for Applied Social Sciences, University of Zimbabwe.

Ndhlovu, Nobuhle, Saito, Osamu, Djalante, Riyanti and Yagi, Nobuyuki. 2017. Assessing the Sensitivity of Small-Scale Fishery Groups to Climate Change in Lake Kariba, Zimbabwe. *Sustainability 9(12)*: 2209.

Nshimbi, Christopher Changwe. 2017. Life in the Fringes: Economic and Sociocultural Practices in the Zambia–Malawi–Mozambique Borderlands in Comparative Perspective. *Journal of Borderlands Studies 34(1)*: 37–70.

Overå, Ragnhild. 2003. Market Development and Investment "Bottlenecks" in the Fisheries of Lake Kariba, Zambia. In Eyolf Jul-Larsen, Jeppe Kolding, Ragnhild Overå, Jesper Raakjær Nielsen, and Paul A.M. van Zwieten (Eds), *Management, Co-Management or No-Management? Major Dilemmas in Southern African Freshwater Fisheries. Part 2: Case Studies*. FAO Fisher. Rome: Food and Agriculture Organization of the United Nations, pp. 201–232.

SADC Secretariat. 2015. Ministerial Migration Dialogue for Southern Africa on: Addressing Mixed and Irregular Migration in the SADC Region: Protection of the Unaccompanied Migrant Child Victoria Falls, Zimbabwe 7–9 of July 2015. In *Addressing Mixed and Irregular Migration in the SADC Region: Protection of the Unaccompanied Migrant Child*. Victoria Falls: SADC Secretariat.

Scudder, Thayer. 1962. *The Ecology of the Gwembe Tonga*. Manchester: Manchester University Press.

Scudder, Thayer. 1968. Social Anthropology, Man-Made Lakes and Population Relocation in Africa. *Anthropological Quarterly 41(3)*: 168–176.

Scudder, Thayer. 2005. *The Kariba Case Study*. California Institute of Technology, Social Science Working Paper 1227, June 2005.

Siamonga, Elliot. 2014. Fishing: Binga Art that Never Dies. *The Patriot Newspaper*, 18 December. www.thepatriot.co.zw/old_posts/fishing-binga-art-that-never-dies/.

Southern African Development Community (SADC). 2001. *SADC Protocol on Fisheries. Protocol on Fisheries*. SADC Secretariat. www.sadc.int/files/8214/7306/3295/ SADC_Protocol_on_Fisheries.pdf.

Southern African Development Community (SADC). 2011. *Regional Strategic Action Plan on Integrated Water Resources Development and Management (2011–2015) RSAP III*. Gaborone, Botswana: SADC.

UN Development Programme. 2017. Real-Time Weather Forecasts Are Helping Zambian Women Farmers Win Their Battle against the Impact of Climate Change – Zambia | ReliefWeb. *Reliefweb.* https://reliefweb.int/report/zambia/real-tim e-weather-forecasts-are-helping-zambian-women-farmers-win-their-battle-against.

Wilson, T. M., and Donnan, H. 1998. Nation, State and Identity at International Borders. In T. M. Wilson and H. Donnan (Eds), *Border Identities: Nation and State at International Frontiers.* Cambridge: Cambridge University Press, pp. 1–30.

Wilson, Thomas M. and Donnan, H. (Eds). 2005. *Culture and Power at the Edges of the State: National Support and Subversion in Europe Border Regions.* Munster: LIT Verlag.

World Fish Center. 2018. Who We Are | WorldFish Organization. www.worldfishcen ter.org/who-we-are.

World Vision. 2003. Zambia: Gwembe Floods Strand Valley Population – Zambia. *ReliefWeb.* https://reliefweb.int/report/zambia/zambia-gwembe-floods-strand-va lley-population.

World Wide Fund For Nature. 2018. Our Global Goals | WWF. WWF Global. http:// wwf.panda.org/what_we_do/how_we_work/our_global_goals/index.cfm.

Xinhua. 2008. Zambia Gov't Describes Hunger Situation in Southern Province as Desperate – Zambia. *Reliefweb.* https://reliefweb.int/report/zambia/zambia-govt-de scribes-hunger-situation-southern-province-desperate.

Zambia Legal Information Institute. 2018. *Fisheries Act, Cap 200.* The Fisheries Act Chapter 200 (Regulation 6): The Fisheries Regulations, Prohibited Fishing Area. www.zambialii.org/zm/legislation/consolidated_act/200.

8 Undocumented migration between Zimbabwe and South Africa
Reflections on migration and peace

Inocent Moyo

Introduction

There is no accepted definition of peace, but for the present purpose I will accept the definition that "peace is about the absence of war, fear, conflict, anxiety, human suffering and violence and about peaceful co-existence" (Francis, 2011: 512). Further, Galtung (1996 cited in Francis, 2011: 512), identifies two types of peace which are "'negative peace' – the absence of direct violence, war, fear and conflict at individual, national, regional and international levels – and 'positive peace' [which is] the absence of unjust structures, unequal relationships and injustice, and inner peace at an individual level". This means that peace is linked to security and development (Francis, 2011). The implication of this is that phenomena such as migration may be taken to be indicative of the absence of peace to the extent that in some cases it may involve human suffering and anxiety, among other things, of the people who migrate, especially undocumented migrants, who are the subject of this chapter.

It is in this context that I reflect on undocumented migration leading to human smuggling between South Africa and Zimbabwe and argue that the absence of war between the two countries does not necessarily mean the existence of peace on the part of the people involved in migration, such as those without immigration documents. This is relevant in a cross-border context between South Africa and Zimbabwe, where migration from the latter to the former has been a common feature of migration in the Southern African region (Wentzel, 2003; Crush et al., 2005). The fact that human smuggling is rampant between nation-states that belong to the same regional economic community – that is, the Southern African Development Community (SADC) – suggests, first that the people involved in the human smuggling process go through painful experiences (or indeed the absence of peace) and second, that the proper management of migration between these two countries could help in creating peace in and for the undocumented migrants. This suggests that achieving peace from this point of view in the SADC region will continue to be a challenge as long as there is no regional migration management protocol in the region. In engaging with these issues, this chapter draws on a qualitative study of a total of 65 Zimbabwean migrants who were interviewed at the Beitbridge border,[1] in Messina[2] and in Johannesburg[3] between December 2014 and March 2015. At the border, 12 migrants were interviewed, of which, three were females and nine were males. At Messina a total number of 34 migrants were interviewed and 19 were women and 15 were men. In Johannesburg, 19 migrants were interviewed of which 13 were men and six were women. The interviews centred on issues relating, but not limited to, reasons for migrating to South Africa, how and why they crossed the border.

Conceptualisation

There is the recognition that migration, peace, security and development are linked, which complicates the process of migration (Mohamed, 2016). With this in mind, there is need to clarify and contextualise the concepts of peace, security and development as they relate to migration in the SADC setting and how they are deployed in this contribution. The definition of peace is hotly contested, but it is generally linked to the absence of violence (Francis, 2011). Violence can be divided into three types, which are "direct violence (physical, emotional and psychological); structural violence (i.e. deliberate policies and structures that cause human suffering, death and harm); and cultural violence (i.e. cultural norms and practices that create discrimination, injustice and human suffering)" (Galtung, 1996, cited in Francis, 2011: 512). This definition of violence means that there are two types of peace, which are positive and negative peace, meaning that peace includes more than the absence of war and incorporates issues of the "inner peace" of individual people (ibid.). The definition of violence and peace, therefore, means that the understanding and interpretation of peace would vary from place to place based on "the particular historical experience and specific political context of a country, society or region" (Francis, 2011: 512). This explains why for instance "the majority of people in the Great Lakes region of Africa associate peace with the absence of war and armed conflict, while those in Southern Africa associate peace with the absence of depressing social and development indicators such as poverty and the AIDS pandemic" (Freedman and Poku, 2005, cited in Francis, 2011: 512).

Despite these differing definitions and interpretations of peace, the point of convergence is that peace in all its manifestations is therefore about security, development and social justice (Francis, 2011: 12). With this in mind, my application of the concept of peace, goes beyond the definition of peace as the absence of war (negative peace) to a discussion of conditions, circumstances and/or policies that affect the individual, even though there may be no war. Hence "peace for 'whom' and for 'what purpose'" (Francis, 2011: 511–512) become important points of reference in a discussion of undocumented migrants from Zimbabwe to South Africa, even if there is no war in or between these two countries. This definition of peace links it to the issue of security, whose definition like that of peace should transcend the focus "on the threat and use of force" (Francis, 2011: 512) to include issues of "survival and the conditions of human existence" (Buzan, 1991; Thomas and Wilkin, 1999, cited in Francis, 2011: 512). There may be "non-military sources of threat to security at the individual, societal, state, regional and global levels", for which cause, the conceptualisation and interpretation of security has to be broadened to illuminate "non-military dimensions such as environment, migration, ethno-religious and nationalist identities, poverty and economic insecurity, and disease" (Francis, 2011: 512). Security, therefore is about the creation and preservation of conditions which make human existence and survival possible, and this means that security is essentially "about peace, development and justice" (Francis, 2011: 512). In this way, it is clear that there is a close relationship between peace and security in the sense that peace and security could be considered to be mutually constitutive, because where and when peace and security prevail, conditions exist for human development to be possible. This also means that development can be seen as a security issue. If development is viewed as human "'progress' or 'change for the better'" there is need for security, hence there have been calls for the "securitisation of development" which means "treating development as a security issue" (Francis, 2011: 512).

From the foregoing, it can be surmised that there is a very close link between peace, security and development because "development is about equality and social justice; positive change at personal, societal, national and global levels; and about peace and security" (Francis, 2011: 513). This close relationship between the three processes is one of the reasons that countries have been pursuing regional economic integration, in the hope that issues of peace and security will be attained, based on the assumption of co-operation, unity and interconnectedness that economic integration yields (Francis, 2011). Concerning regional integration and co-operation, the SADC works towards, among others, economic progress, peace and security (SADC, 1992). However, it appears the treatment of the issue of peace and security is limited to armed conflict and war. Accordingly, this chapter debates and reflects on the implications for peace of the way that undocumented migrants who move from one SADC country to another are treated and by these means engage with the concept of peace, security and development, from the point of view of the migrating or moving body in the form of the undocumented migrant. This speaks to those policies in many countries that deter and create a border against migrants because of the perception that they threaten peace and security in the destination countries (Riera-Cézanne, 2017). In the case of Europe this has led to the erection of barriers against migrants especially those from Africa and parts of the Middle East (Nshimbi and Moyo, 2016). In this chapter, I posit that such policies actually assault the peace, security and development of such migrants, even though they are assumed to uphold and protect the same.

Undocumented migration, human smuggling, *impisi* and *omagumaguma*

Although migration from Zimbabwe to South Africa has been going on for a longer time, two phases of migration between these two countries can be identified in the period since Zimbabwe's independence in 1980 (Tevera and Zinyama, 2002). Soon after independence in 1980, white skilled and semi-skilled workers left Zimbabwe for many countries, including South Africa; but the recent economic collapse and political instability resulted in renewed flows of people from Zimbabwe to South Africa (Tevera and Zinyama, 2002). For instance, in the early 1990s the introduction of the Economic and Structural Adjustment Programme (ESAP) led to deteriorating economic conditions (Tevera and Zinyama, 2002). This is because the neoliberal economic model on which ESAP was based weakened the economy leading to declining economic growth, which led to the closure of industries, retrenchments and unemployment (Dhemba, 1999; Brett, 2005). More recently, the negative economic performance in Zimbabwe has been linked to the land reform programme (Raftopoulos, 2006). It is argued that the land reform programme affected agro-industries which resulted in a negative investment climate, unemployment and general economic malaise and paralysis (Fontein, 2009).

It needs to be mentioned that Zimbabweans who migrate to South Africa include those who are documented and those who are not. My focus in this chapter is on the undocumented category, because they illuminate the subject of this contribution, which is the lack of peace and how it may be difficult to achieve. This is because undocumented migration leads to human smuggling. At the Beitbridge border post, Araia (2009), notes that human smuggling is constituted of many processes which include: crossing the Limpopo River;[4] scaling the border fence; and crossing the official border without travel documents but with the assistance of some immigration officials. This involves many key players such as: taxi operators (*omalayitsha*), government officials and other migrants (Araia, 2009). For example, Taxi[5] operators transport individuals from different parts of Zimbabwe to selected points in Beitbridge town. When the number of migrants who are transported by different taxi operators is substantial (30–40 people) they go to specific crossing points from where they are taken across the border by guides (*impisi*) who are familiar with the area and are hired by the taxi operators (Interview with a taxi operator/*malayitsha*, Beitbridge, 4 January 2015).

The *impisi* generally ensure that undocumented migrants cannot be easily arrested by the South African Police Service (SAPS) and/or South African National Defence Force (SANDF) at illegal crossing points along the Limpopo River and the border fence (Interview with a taxi operator/*malayitsha*, Beitbridge, 4 January 2015). Further, some government officials from both the Zimbabwean and South African sides assist undocumented Zimbabweans to cross into South Africa at the Beitbridge border post by accepting bribes (Araia, 2009). After undocumented migrants have crossed the border, the taxi operators fetch them from designated points along the freeway from Beit-bridge to Messina (ibid.).

Interview data seems to show that sometimes undocumented migrants go through traumatic experiences and some meet painful deaths before those who survive board taxis to Johannesburg. This is because of the presence of a violent gang of criminals (*omagumaguma*), who rob the undocumented migrants of their valuables and also rape and kill women (Interview with a Zimbabwean migrant, Messina, 5 January 2015). There is no agreement on the nature and dynamic of *omagumaguma*, because some studies suggest that they assist undocumented Zimbabweans to cross into South Africa, but in the process rob and sometimes kill such undocumented migrants (Ndlovu, 2013: 1151), which means *impisi* and *omagumaguma* can be the same person (Araia, 2009). Whether or not *omagumaguma* operate independently of *impisi* or *omalayitsha* (Mdlongwa and Moyo, 2014), they are clearly "bandits who have been attracted to the banks of Limpopo to prey upon migrants" (UN Office for the Coordination of Humanitarian Affairs, 2010: 19). What is even more instructive is that *omagumaguma* are also involved in the lucrative business of smuggling goods such as tobacco from Zimbabwe into South Africa (Nqindi, 2012). Important for this chapter are the experiences of the undocumented Zimbabwean migrants from the perspective of peace outlined in the beginning of this contribution. Hence the question of how the experiences of undocumented Zimbabwean migrants demonstrates the lack of peace even if there is no war, in Zimbabwe or South Africa or between Zimbabwe and South Africa comes into sharp focus. To provide a context for a discussion of this, it is needful to consider how the undocumented Zimbabwean migrants are treated in South Africa after the smuggling experiences discussed in this section. Against this background, a brief discussion of the temporary dispensation, special and exemption permit regimes extended by the South African government to undocumented Zimbabwean migrants is necessary.

Dispensation of Zimbabweans Project (DZP), Zimbabwean Special Dispensation Permits (ZSP) and Zimbabwean Exemption Permits (ZEP)

Since 2009, the South African government has attempted to document Zimbabweans in South Africa through a series of temporary permits. For instance, in 2009, the South African government approved permits for Zimbabweans under the DZP project. This allowed Zimbabweans to work, conduct business and study in South Africa (Gigaba, 2014). The objectives of the DZP permits were: to regularise undocumented Zimbabweans; curb the deportation of undocumented Zimbabweans; reduce pressure on the asylum and refugee system; and provide an amnesty to Zimbabweans who had obtained fraudulent South African documents (Gigaba, 2014). Approximately 295,000 Zimbabweans applied for DZP permits of which approximately 245,000 were issued (Gigaba, 2014). As the DZP permits were due to expire at the end of December 2014, the Minister of Home Affairs (Malusi Gigaba) announced on 12 August 2014 that these permits would be replaced by ZSP documents. The ZSP are similar to the DZP in that they were temporary residence permits which allowed Zimbabweans to live, work, conduct business and study in South Africa. They expired at the end of 2017 (Gigaba, 2014). The Minister of Home Affairs noted that "the ZSP is a temporary bridge to the near future when all Zimbabweans will re-enter the mainstream immigration process in South Africa" (Gigaba, 2014).

But, unlike the DZP, there are several conditions attached to the ZSPs, such as that they are non-renewable and the permit holder does not qualify for permanent residence on the basis of the time spent in South Africa (Interview with a ZSP permit holder, Johannesburg, 28 February 2015). However, in September 2017, the Minister of Home Affairs Professor Hlengiwe Mkhize announced that the ZSP permits will be replaced by a new category of permits called Zimbabwean Exemption Permits (ZEP), for four years up to December 2021. The ZEP entitle holders to work or conduct business in South Africa, but their holders are not entitled to apply for permanent residence. Further, these permits are not renewable or extendable and their conditions cannot be changed as long as the holder is in South Africa (Mkhize, 2017).

Two issues are worth highlighting at this stage. First, some undocumented Zimbabwean migrants die in their quest to migrate to South Africa, as they either drown in the Limpopo River or are eaten by crocodiles in the same river, or indeed killed by gangs in the area (Interview with a taxi operator/ *malayitsha*, Beitbridge, 4 January 2015). Second, those who survive and manage to get to South Africa are placed on a series of temporary permits with limited rights. This chapter will comment on the implications of this in terms of the peace of the migrants involved.

Migration experiences and peace

In an attempt to explore the connection between undocumented migration and peace regarding Zimbabweans attempting to migrate to South Africa, this discussion will dwell on two issues. The first is the migration experiences of the undocumented as they went through the human smuggling process. The second is how they were treated when they arrived in South Africa. Concerning the migration experiences of undocumented Zimbabwean migrants, it can be noted that they had traumatic experiences. This is in the form of evading crocodiles in the Limpopo River, scaling the fence and being led through the bush by *impisi*. Some of the migrants are eaten by crocodiles and yet others are killed by *omagumaguma*. Those who survive should be scared, as a result of going through near-death experiences and witnessing some of their fellow Zimbabweans being eaten by crocodiles or raped and killed by *omagumaguma*. This suggests that such Zimbabwean migrants do not enjoy peace even if there is no actual war.

The views of Zimbabweans concerning some of the temporary permit programmes provide an insight into how such migrants are troubled. Although the ZSP programmes will be replaced by the ZEP, comments by Zimbabwean migrants make important points of reference. A case in point relates to the announcement concerning ZEP, that holders of the permits cannot change the conditions of their permits while in South Africa (suggesting that they must go back to Zimbabwe at the expiration of the permits) and they cannot apply for permanent residence, despite the fact that they would have officially lived in South Africa for over a decade. Respondents noted that there was no life (they had left their jobs and had set up a new life in South Africa) for them in Zimbabwe as they had been in South Africa for a long period of time (Interview with a ZSP permit holder, Johannesburg, 28 February 2015). These Zimbabweans feel that the requirement to apply for South African immigration visas from Zimbabwe is tantamount to "chasing them from South Africa", for which reason they were prepared to find other means to remain in South Africa (Interview with a ZSP permit holder, Johannesburg, 28 February 2015, see Moyo, 2018 for a detailed discussion).

The important point is that even as Zimbabweans are required to go back to and apply for mainstream permits from Zimbabwe, there is no guarantee that they will be granted these. The implication of this is that such Zimbabweans must stay in Zimbabwe, because when they migrate to South Africa as undocumented migrants, they will face the full wrath of the law in terms of arrest and detention. This seems to demonstrate two things. First is the mobilisation of the humanitarian logic and second is the depoliticisation of these migrants (Moyo, 2018). Concerning the former, it "designate(s) the deployment of moral sentiments in contemporary politics", it puts in place "actions conducted in order to manage, regulate and support the existence of human beings" (Fassin, 2012: 1). At the centre of the humanitarian government "is a new moral economy of suffering [...] this new moral economy represents not simply a way of relating to human suffering, but a mode of governing it" (Reid-Henry, 2013: 756). On this basis, it would appear as if the permits in question illustrate:

> a form of cynicism at play when one deploys the language of moral sentiments at the same time as implementing policies that increase social inequality, measures that restrict the rights of immigrant populations ... in this view, the language of humanitarianism would be no more than a smoke screen that plays on sentiment in order to impose the law of the market and the brutality of realpolitik.
>
> (Fassin, 2012: 2)

This is because the permits that are intended to document Zimbabweans have within them conditions that actually place the beneficiaries in a tenuous immigration status.

When Zimbabweans who have stayed in South Africa for over ten years are told that they cannot qualify and apply for permanent residence but must go back to Zimbabwe and apply for permits that they are not guaranteed to get, this actually limits the rights of such migrants and increases inequality between them and the citizens. In this way, "humanitarianism has become the language that inextricably links values and affects, and serves both to define and to justify discourses and practices of the government of human beings" (Fassin, 2012: 2). The conditions of these permits have led to some Zimbabwean migrants who hold ZSP documents regretting surrendering their fraudulently acquired South African documents in order to apply for the DZP and ZSP documents (Interview with a ZSP permit holder, Johannesburg, 28 February 2015). This shows how troubled such migrants are, because of the feeling that the dispensation, special and exemption permits are failing on their key objective of documenting Zimbabweans.

Beyond the condition of the DZP, ZSP and ZEP, the mere fact of placing Zimbabwean migrants on these temporary permits for longer periods of time without a guarantee for either permanent residence or the other so-called mainstream permits actually depoliticises and *invisibilises* such migrants. This is because they are framed as visitors who must go back to Zimbabwe (Moyo, 2018). Differently stated, "they are permitted, but not necessarily welcome" (Reid-Henry, 2013: 756). They are permitted on the basis of an extension of a humanitarian favour and when it is withdrawn, those from whom it has been withdrawn have no basis for complaining, because "those at the receiving end of humanitarian attention know quite well that they are expected to show the humility of the beholden rather than express demands for rights" (Fassin, 2012: 4). Within this logic, the Zimbabwean migrants are depoliticised and *invisibilised* and this places them in a state of liminality (Moyo, 2018). They are not certain of what the future holds for them given that they cannot apply for permanent residence and there is no guarantee they will be granted the mainstream permits (Moyo, 2018).

Managing migration to achieve peace, security and development: some concluding remarks

The history as well as the current trends in migration from especially Zimbabwe to South Africa seem to suggest that this is a permanent reality, which may be difficult to stop (Moyo, 2017; Moyo and Nshimbi, 2017; Nshimbi and Moyo, 2017). The implication is that the difficult conditions through which migrants are smuggled and for those who make it to South Africa to exist in a tenuous immigration status are most likely to continue as well. It is accepted that Zimbabwe is responsible for uprooting its own citizens such that they migrate to countries like South Africa, seeking, among other things, peace, security and personal development. It is in this context that their treatment in the destination country comes to the fore.

The reasons that the migrants in question gave for choosing to be smuggled centre on their being not able to access the necessary South African immigration documents and for those who do, the conditions on these documents place them in a state of liminality. A case in point relates to the DZP, ZSP and ZEP permits, which have many conditions. These conditions limit the rights of their holders such as the non-renewability clause and the fact that their recipients cannot apply for permanent residence. From the point of view of the definition of peace which guides this chapter, such Zimbabweans lack security and development in South Africa. The fact that they do not know whether or not their permits will be renewed and that they cannot apply for permanent residence should be seen in light of the fact that most of the respondents stated that there was no life for them in Zimbabwe. The affected migrants therefore have no peace, security and personal development. They cannot make future plans in South Africa because they are expected to go back to Zimbabwe and they cannot properly plan their future in Zimbabwe because there are a host of challenges which made them migrate to South Africa in the first instance (Moyo, 2018). Personal development in such circumstances is difficult. It is difficult for such migrants to plan for an uncertain future either in South Africa or Zimbabwe and this illuminates the lack of peace, security and development. Going forward, there is need for proactive strategies for managing migration between SADC member states such as that between Zimbabwe and South Africa. It is acknowledged that the DZP, ZSP and ZEP permits were and still are attempts to address the unique situation of Zimbabwean migrants. However, the exclusionary conditions attached to such permits seem to contradict its primary aim of documenting Zimbabweans and assisting them to work, earn a living and thus achieve security and development, which they have failed to do in Zimbabwe – the reason why they migrated to South Africa in the first instance (Moyo, 2018). In case such Zimbabweans are not granted mainstream permits at the expiration of their permits, and the application for permanent residence is out of the way, this implies that such Zimbabweans may actually engage in human smuggling to come to South Africa, where they have built or are attempting to build a future (Moyo, 2018). In this, I see that migration, security and development are inextricably linked concerning the case of undocumented migrants from Zimbabwe to South Africa.

Nonetheless, there is a foundation on which countries like South Africa can build so as to contribute to the peace, security and development of undocumented migrants such as those from Zimbabwe. This foundation is the DZP, ZSP and ZEP permits that the South African government has extended to Zimbabweans. This means that the South African government should consider relaxing conditions on these permits in an attempt to amplify and indeed contribute to the aspirations and achievement of the goals of the SADC Draft Protocol on the Facilitation of Movement of Persons (Moyo, 2018). The overarching aim of this Protocol is to develop policies that will progressively eliminate obstacles to human mobility into and within SADC

Members' territories (SADC, 2005: 3) in line with the Declaration and Treaty of SADC (SADC, 1992) concerning human mobility in the region. As an SADC member state, South Africa understands that the regional block aims "to strengthen and consolidate the long standing historical, social and cultural affinities and links among the people of the region" (SADC, 1992: 5).

Pursuant to this, the SADC Declaration and Treaty of 1992, Article 5.2 (d) commits SADC countries to implementing "policies aimed at the progressive elimination of obstacles to the free movement of capital and labour, goods and services, and of the people of the region generally, among Member States" (SADC, 1992: 5). Therefore, strengthening the current temporary permits extended to Zimbabweans by relaxing the conditions and expanding the rights thereof, may reduce human smuggling, leading to free movement (Moyo, 2018). This free movement would prevent the need to irregularly cross the border and work in South Africa without correct documentation. This would contribute to peace as it relates to the migrants themselves. They are uprooted from Zimbabwe as a result of a host of vulnerabilities, they suffer the same as they cross the border and attempt to earn a livelihood in South Africa. This means that there is no peace for the migrants where they come from, on their journey or in their destination area. This assaults the very idea of regional integration, peace, security and development.

Notes

1 Beitbridge border post is the official entry and exit point between Zimbabwe and South Africa.
2 Messina is a South African town situated approximated 17 kilometres from the Beitbridge border.
3 Johannesburg is the largest city of South Africa, situated approximately 521 kilometres from the Beitbridge border post.
4 The Limpopo River forms a natural boundary between Limpopo Provence, in Northern South Africa and Southern Zimbabwe.
5 Taxi here refers to a form of public transport that is commonly referred to as kombis in the region. They are similar to minibuses with the capacity to carry 15–20 people.

References

Araia, T. 2009. *Report on Human Smuggling across the South Africa/Beitbridge Border.* MRMP Occasional Paper. Johannesburg: Forced Migration Studies Programme, University of Witwatersrand.

Benhabib, S. 2014. Critique of Humanitarian Reason. *Eurozine.* www.eurozine.com/critique-of-humanitarian-reason/.

Brett, E. A. 2005. From Corporatism to Liberalisation in Zimbabwe: Economic Policy Regimes and Political Crisis, 1980–1997. *Journal of International Political Science Review* 26(1): 91–106.

Crush, J., Williams, V. and Peberdy, S. 2005. Migration in Southern Africa. Paper prepared for Policy analysis and Research Programme of the Global Commission on International Migration. September 2005.

Crush, J. and Tevera, D. 2010. Exiting Zimbabwe. In J. Crush and D. Tevera (Eds), *Zimbabwe's Exodus: Crisis, Migration, Survival*. Kingston and Cape Town: Southern African Migration Programme, pp. 1–49.

Dhemba, J. 1999. Informal Sector Development: A Strategy for Alleviating Urban Poverty in Zimbabwe. *Journal of Social Development in Africa* 14(2): 5–19.

Fassin, D. 2012. *Humanitarian Reason: A Moral History of the Present*. Berkeley: University of California Press.

Fontein, J. 2009. Anticipating the Tsunami: Rumours, Planning and the Arbitrary State in Zimbabwe. *Africa* 79(3): 370–398.

Francis, T. 2011. Linking Peace, Security and Developmental Regionalism: Regional Economic and Security Integration in Africa. In Erin McCandless and Tony Karbo (Eds), *Peace, Conflict and Development in Africa: A Reader*. Geneva, Switzerland: University for Peace, pp. 210–520.

Gasa, N. 2009. Tell Them We Are Not Dead Yet. *The Star*, 24 February, p. 11.

Gigaba, M. 2014. DZP for Zimbabweans in SA to be Replaced by ZSP. *politicsweb*. www.politicsweb.co.za/party/dzp-for-zimbabweans-in-sa-to-be-replaced-by-zsp-m.

Hammerstad, A. 2011. *Linking South Africa's Immigration Policy and Zimbabwe Diplomacy*. SAIIA Policy Briefing 42. Johannesburg: SAIIA.

International Organization for Migration (IOM). 2010. *Wolves in Sheep's Skin: A Rapid Assessment of Human Trafficking at Messina, Limpopo Province of South Africa*. Geneva: International Organization for Migration.

Mdlongwa, T. and Moyo, T. 2014. Representations of Xenophobic Otherisation in Jinga's One Foreigner's ordeal and Mpe's Welcome to our Hillbrow. *Elite Research Journal of Education and Review* 2(4): 88–94.

Mkhize, H. B. 2017. Statement by minister Mkhize on the closure of the Zimbabwean Special Permits (ZSP) and the opening of the New Zimbabwean Exemption Permits (ZEP). www.dha.gov.za/index.php/statements-speeches/1034-statement-by-minister-mkhize-on-the-closure-of-the-zimbabwean-special-permit-zsp-and-the-opening-of-the-new-zimbabwean-exemption-permit-zep (accessed 13 October 2017).Mlambo, E. 2014. Renewed Exodus Hits Zimbabwe. *Zimbabwe Independent*. www.theindependent.co.zw/2014/05/02/renewed-exodus-hits-zimbabwe/.

Mohamed, N. 2016. Migration, Peace and Development: When Concrete Action Clashes with a Positive Vision. Oxfam. http://oxfameu.blogactiv.eu/2016/12/22/migration-peace-and-development-when-concrete-action-clashes-with-a-positive-vision/.

Moyo, I. 2017. Zimbabwean Cross-Border Traders in Botswana and South Africa: Perspectives on SADC Regional Integration. In C. C. Nshimbi and I. Moyo (Eds), *Migration, Cross-Border Trade and Development in Africa: Exploring the Role of Non-state Actors in the SADC Region*. London: Palgrave, pp. 43–62.

Moyo, I. 2018. Zimbabwean Dispensation, Special and Exemption Permits in South Africa: On Humanitarian Logic, Depoliticisation and Invisibilisation of Migrants. *Journal of Asian and African Studies*. doi.org/doi:10.1177/0021909618776413.

Moyo, I. and Nshimbi, C. C. 2017. Cross Border Movements and Trade: A Tenacious Lasting Reality in the Southern African Development Community (SADC) Region. In C. C. Nshimbi and I. Moyo (Eds), *Migration, Cross-Border Trade and Development in Africa: Exploring the Role of Non-state Actors in the SADC Region*. London: Palgrave, pp. 191–208.

Ndlovu, L. 2013. They Are Everywhere Like Cockroaches: Unpacking the Experiences of the Zimbabweans in the Diaspora through the Short Essay. *European Journal of Humanities and Social Sciences* 22(1): 1143–1158.

Nqindi, J. 2012. Zimbabwe's Illicit Smokes Flow into South Africa. *Sunday Times*, p. 8.

Nshimbi, C. C. and Moyo, I. 2016. Visible and Invisible Bordering Practices: the EU–African Migration Conundrum and Spatial Mobility of Borders. *World Journal of Science, Technology and Sustainable Development* 13(4): 300–314.

NshimbiC. C. and MoyoI. 2017. History, Trends and Dynamics of Cross Border Movements and Trade in the Southern African Development Community (SADC) Region. In C. C. Nshimbi and I. Moyo (Eds), *Migration, Cross-Border Trade and Development in Africa: Exploring the Role of Non-state Actors in the SADC Region.* London: Palgrave, pp. 1–14.

Raftopoulos, B. 2006. The Zimbabwean Crisis and the Challenges for the Left. *Journal of Southern African Studies* 32(2): 203–209.

Reid-Henry, S. 2013. Review Essay: On the Politics of our Humanitarian Present. *Environment and Planning D: Society and Space* 31: 753–760.

Riera-Cézanne, J. 2017. Migration and Its Link to Peace, Security, and Sustainable Development Agenda. https://reliefweb.int/training/2250319/migration-and-its-link-peace-security-and-sustainable-development-agenda (accessed 11 November 2017).

Scheen, T. 2011. Zimbabwean Migrants Destabilise the North of South Africa. *Rural* 21(1): 16–17.

Seale, L. and Tromp, B. 2009. While Politicians Fiddle a Zimbabwean Man Dies. *The Star*, 24 February, p. 3.

Southern African Development Community (SADC). 1992. SADC Declaration of Treaty. www.sadc.int/documents-publications/sadc-treaty/.

Southern African Development Community (SADC). 2005. Draft Protocol on the Facilitation of Movement of Persons. www.sadc.int/documents-publications/show/Protocol_on_Facilitation_of_Movement_of_Persons2005.pdf.

Tevera, D. and Zinyama, L. 2002. *Zimbabweans Who Migrate: Perspectives on International Migration in Zimbabwe.* Migration Policy Series No 2. Cape Town: IDASA.

United Nations Office for the Coordination of Humanitarian Affairs. 2010. ochadms. unog.ch/.../04.5.2.1_Sector_Response_MAP_Food Zimbabwe_10062015 (accessed 16 April 2016).

United Nations Office on Drugs and Crime (UNODC). 2009. *UNODC Activities in Support of the Implementation of the Trafficking in Persons and Smuggling of Migrants Protocols.* Washington, DC: United Nations.

Wentzel, M. 2003. Historical and Contemporary Dimensions of Migration between South Africa and its Neighbouring Countries. Unpublished paper delivered at HSRC migration workshop, Pretoria, 17–20 March 2003.

9   An examination of the SADC regional integration posture in the context of the contested 2008 election run-off in Zimbabwe

Patrick Dzimiri

Introduction

Globally, the role of elections in the entrenchment of democracy is increasingly appreciating in value. Africa is not missing out on the processes as almost all the countries have embraced the conduct of elections in legitimising their leadership. The conduct of elections is also consistent with Agenda 2063, which is Africa's 2063 transformative agenda, and Vision 2030 of the United Nations (UN) regarding sustainable development, peace, security and development. Just in 2017 alone, about four African countries, namely Liberia, Rwanda, Angola and Kenya, conducted elections as a fulfillment of the democratisation requirements. The realisation that multiparty democracy and the holding of elections is the legitimate process for assuming power has also translated to increased competition for power, with severe costs. One major set-back in the electoral processes in Africa is that, since the 1990s, violence has constantly featured in and characterised the conduct of elections. There is also a view that the entrenchment of democracy in Africa is being perverted by governments established on revolutionary foundations (Guichaoua, 2017). Most of the revolutionary regimes fail to manage opposition political parties and they rely upon military muscle to stay in power. This has been the case in Zimbabwe, Kenya, Angola, Namibia, Rwanda and the Democratic Republic of Congo (DRC), among others (Bekoe, 2010).

Another observable and recurring phenomenon in the conducting of elections is that the outcome is often predetermined, and this is attributed to authoritarian governance and lack of independence on the part of election management bodies (EMBs). The 2016, DRC and 2017, Burundian experiences show limitless pursuit of power by African leaders. Leadership in these two countries attempted to manipulate the constitution to prolong their stay in power. As rightly noted by Guichaoua (2017), elections in many parts of Africa are therefore increasingly becoming a mere political ritual as they are neither free nor fair. The outcome is always contested and violence is usually perpetrated by the incumbent regime as a strategy for imposing the election outcome (Collier and Vicente, 2012). This also corresponds with the remark by Motsamai (2010: 3) that "violence also ensues in situations where there is a strong possibility of changing existing [political order] and when the incumbents are unwilling to cede power". This may be the reason why in some of the elections, the outcome is never decisive and is highly contested. This has been the case in Zimbabwe in 2008 and Kenya in 2017, where election run-off had to be instituted.

More disquieting is that, where violence has manifested in elections as in the case of Zimbabwe, the socio-economic and political results of such, have transcended the frontiers of the state. A study conducted by Gerenge (2015) attests to this observation, citing the cases of Zimbabwe in 2008, Lesotho in 1998, Kenya in 2007, DRC in 2011 and Ivory Coast in 2010 where election related humanitarian tragedy and violence spilled over into neighbouring countries. In most of these cases, occurrence of pre- and post-election violence has been attributed to parties defaulting or not complying with international and regional norms as well as the agreed upon standards that govern the conduct of democratic elections (Gerenge, 2015). Reading from these sad realities of elections in Africa, one would be vindicated in saying that elections are not a panacea for democratisation and stability in Africa. In light of the above preliminary discussions, this chapter unpacks the causes, patterns and contexts of the 2008 election-related violence in Zimbabwe to demonstrate that electoral conflicts have spillover implications for regional peace and security as well as integration. South Africa and Botswana, for example, were and are still the most afflicted by the crisis. Another salient revelation by the study is that women were highly affected by the 2008 election violence and many of them had to risk their lives in cross-border journeys as they tried to fend for their families. The chapter concludes by revealing that, for purposes of preserving SADC regional integration, state-centred approaches were invoked in resolving the electoral crisis in Zimbabwe.

Conceptual considerations

From power and governance's perspectives, elections are increasingly becoming a peaceful method available to political parties or groups to openly compete for power and for citizens to legitimately elect their representatives or leadership for purposes of effective governance (Wallsworth, 2015; Brown, 2003). The challenge, however, is that the process of democratisation through regular elections in the context of Africa has some serious human security costs. Since 1990, African states have been holding elections consistently, but this has not transformed into representative democracy. Gloomy electoral outcomes in Africa parallel the postulation by Lindberg (2009: 328) that "the link between elections and democratisation is not theoretically tied to freedom and fairness of elections". This disappointing connection between elections and democracy gave impetus to the notion of the "fallacy of electoralism" (Lindberg, 2009: 328). The reasoning is that, instead of facilitating democratisation, the reality is that elections can be a source of violent conflict. This has been the case in certain parts of Africa such as Zimbabwe, DRC, Kenya and Ivory Coast, among others.

The fact that most elections are marred by violence and have serious human security ramifications, makes electoral violence a unique and deserving area of study. The broad definition of election violence postulated by Fischer (2002) considers any random or organised act or threat to intimidate, physically harm, blackmail or abuse a political stakeholder in seeking to determine, delay or to otherwise influence an electoral process. Straus and Taylor (2009) further elaborate on the manifestation of electoral violence, citing acts such as beatings, torture, murder, rape, forced disappearance, displacement of people and other forms of coercive means. What these definitions suggest is that perpetrators of violence strategically use violence to decide or confuse election outcomes. The nature of the regime is therefore an important unit of analysis when examining elections and democracy. As alluded to by Fisher (2013), countries with unconsolidated democracy or authoritarian alignments have an inclination to use ethnic, tribal and religious cleavages in fuelling violence, as in the case of Kenya (Fisher, 2013). In the case of Zimbabwe, the militarisation of politics as well as politicisation of the unemployed youth have featured in several typologies as the prime cause for electoral violence (Mapuva and Muyengwa-Mapuva, 2014; Chitiyo, 2009).

Electoral violence in one country can beget violence in neighbouring countries. This is apparent in the context of globalisation characterised by regional integration. Regional integration is pursued as an essential force for economic and political cooperation (Mapuva and Muyengwa-Mapuva, 2014; Chingono and Nakana, 2009). The understanding is that through cooperation and integration, countries can collectively overcome various global problems confronting them. Other perceived dividends that regional integration creates are larger economic opportunities, which allows for economies of scale and enhancement of efficiency and growth (Barnekow and Kulkarni, 2017; Mapuva and Muyengwa-Mapuva, 2014; Chingono and Nakana, 2009). In the context of Africa, economic and political vulnerabilities compelled states to resort to regional integration to strengthen bargaining power and stature on the international scene. The above perspectives on regional integration resonate with the formation and existence of the Economic Community of West African States (ECOWAS), the Common Market for Eastern and Southern African States (COMESA) and the Southern African Development Community (SADC) among others in the context of Africa. Conceptually, at the political level, integration connotes normative and institutional dimensions.

Normatively, it involves establishment of common legal, political values, while institutionally it envisages strengthening of democratic institutions and the maintenance of peace and stability for participating countries and citizens (Chingono and Nakana, 2009). In the context of SADC, the Protocol on Elections and Democracy is a typical example of normative benchmarks, while the 1993 SADC Organ on Politics Defense and Security (OPDS) constitutes the institutional set-up or common security regime for promotion of peace and security. This illuminates that security is an integral component of regional cooperation and integration.

A brief history of SADC will make this point clearer. SADC as a sub-regional bloc owes its establishment to colonialism and wars of liberation struggle. Shared history of colonialism and political experiences prompted independent Southern African states in the 1970s, namely, Zambia, Botswana, Mozambique, Angola and Tanzania to come up with an organisation called the Front-Line States (FLS). These countries joined forces in fighting and supporting liberation wars and all forms of resistance to colonial rule in Zimbabwe, Namibia and South Africa (Adolfo, 2009). With the attainment of independence in Zimbabwe in 1980, the FLS transformed into the Southern African Development Coordination Conference (SADCC). The transformation was instituted to meet the economic and political needs of the region as well as reduce economic dependence on apartheid South Africa (Landsberg, 2013; Schoeman, 2009).

With Zimbabwe coming onboard, SADCC geared itself to counter destructive engagements perpetrated by apartheid South Africa on the region. The economic and political independence nexus was further broadened in 1992 when SADCC transformed to the Southern African Development Community (SADC). In the context of the effects of destructive wars of liberation, both socially and economically, SADC had to refocus itself in terms of economic and development cooperation. This is well captured in the Declaration and Treaty of SADC which underlines promotion of economic and social development through cooperation and integration (Declaration and Treaty of SADC – Art 5 (1), 1992). Based on the above vision and aspirations, the SADC regional integration has been premised on cooperation and peaceful inter-state conduct and these settled norms have stood the test of time.

Furthermore, SADC's political, peace and security priorities are enshrined within the Strategic Indicative Plan for the Organ (SIPO) that was adopted by heads of states in 2004 as the strategic framework for SADC's operation in the region (Landsberg, 2013). In a nutshell, the 1992 SADC transformation made it both an economic and security cluster, hence the notion of a security community (Ngoma, 2003). As propounded by Wendt (1995) the idea of security community denotes a social structure comprised of shared knowledge and trust in the resolution of disputes and conflicts. Taylor (2014) augments that shared norms, common understanding and ideas are integral to the achievement of a secure community. In the SADC case, this is expressly covered by the mutual defence pact, namely the Organ on Politics, Defense and Security (OPDS). Chingono and Nakana (2009) expand on the same point of mutual co-existence, arguing that SADC regional integration derives insights from the observation that neighbouring countries that have common social, economic, political and security challenges can survive through mutual interdependence. As such, the SADC Regional Indicative Strategic and Development Plan (RISDP) was crafted as a manual for regional integration while the OPDS was instituted for promoting peace and security in the region.

As an appreciation of the importance of elections for democratic entrenchment in the region, SADC crafted some guidelines for best practices. These include the SADC Principles Guidelines Governing Democratic Elections of 2004. The challenge so far however is that with no enforcement mechanisms in place in the event of non-adherence by member states, the guidelines remain a political rhetoric. As will be shown in the following discussion on the electoral conflict in Zimbabwe, election guidelines are further rendered impotent by the fact that the guidelines call for member countries to resolve election conflicts in accordance with their domestic laws.

The 2008 Zimbabwean electoral conflict in the SADC regional context

Understanding the socio-political environment leading to SADC's involvement in Zimbabwe's political conundrum is important for examining the regional dimensions of the 2008 electoral conflict. Scholarly writings as well as official SADC communiqués show that since 2000, Zimbabwe has been on the SADC radar due to its political conflict ranging from the land reform crisis to electoral politics. Initially, deteriorating human security conditions following the violent land grabs compelled SADC to establish the SADC Task Force Team to engage with all the political parties in the country (Mhandara and Pooe, 2013; SADC Communiqué, 2000). At that time, no robust measures were instituted since the regional leadership were hopeful that parties to the land conflict would appeal to the force of logic and resolve the conflict cordially. More importantly, one would be inclined to assert that there was no urgency in addressing the crisis since it involved the white settler community whom they believed were villains.

The March 2007 pre-election clash between ZANU-PF and the MDC, however, marked SADC's critical entry into the electoral politics of Zimbabwe. This followed the 11 March 2007, Save Zimbabwe Campaign, which was comprised of faith-based organisations, civic organisations and trade unions and the leadership of the MDC. The Public Order and Security Act (POSA), as part of the respective pieces of legislation was invoked to criminalise the peaceful march/gathering that was intended to raise awareness on the deteriorating human rights and economic situation in the country (Mlambo and Raftopoulos, 2010; Bratton and Masunungure, 2008). The excessive use of force by the police to dismantle the gathering resulted in daylight assault on MDC-T Leader Mr Morgen Tsvangirayi and other dignitaries. This unprecedented attack on the opposition leader generated massive international pressure on the SADC leadership to act.

Consequently, the regional leadership were forced to convene an urgent meeting in Tanzania for purposes of addressing looming electoral conflict in Zimbabwe (Murithi and Mawadza, 2011). The outcome of the Tanzania SADC summit endorsed then South African President Thabo Mbeki to lead the SADC facilitation team in the political conflict in Zimbabwe (Murithi and Mawadza, 2011). Mbeki's selection as mediator emanated from the fact that since 2000 he had been involved in the political developments in Zimbabwe, and hence was more informed (Bratton and Masunungure, 2008). In addition to Mbeki's familiarity with the political complexities in Zimbabwe, it would seem like the experiences of election violence during the 2002 presidential and 2005 parliamentary elections sent some early warning signals on the potential destabilisation to the region that may be caused by inaction towards the political developments in Zimbabwe. To appreciate SADC's involvement in Zimbabwe there is need for critical profiling of the political conduct and electoral outcomes of the country since independence in 1980. More glaring is the instrumental role of violence in mobilisation for political support.

The ruling party, ZANU-PF's strategy of using violence for electoral ends has been an unsettling phenomenon since independence in 1980 (Makumbe, 2002). As a result, the political space remained largely restricted and no opposition political party has enjoyed partaking in electoral affairs of the country. This may also be attributed to some colonial antecedences where ZANU-PF adopted the authoritarian conduct of the colonial government and failed to transform and democratise its political conduct (Muzondidya, 2009). Against the continued dominance by ZANU-PF for almost two decades (a *de facto* one party system), the emergence of the MDC as an equal power contender was not a welcome development for the ruling party. The MDC's association with civic organisations and workers' unions threatened ZANU-PF's *de facto* one party model and this was unprecedented. ZANU-PF as a ruling party responded to these tides of change by restricting the political space. The promulgation of repressive pieces of legislation such as the Public Order and Security Act (POSA) and the Access to Information and Protection of Privacy Act (AIPPA) are noticeable examples. It also explains why since the formation of the MDC in 1999, the pre- and post-2000 election environments in Zimbabwe have experienced varying degrees of electoral related violence. Violence at the 2002 presidential elections and the 2005 parliamentary elections are living testimonies. From a regional integration outlook, this also translates to saying that SADC as a regional organisation has been complicit over the electoral crisis in Zimbabwe.

With this piece of history, the task of the SADC facilitation team was, therefore, primarily to create a conducive environment for undisputed elections. A milestone achievement by the Mbeki-led SADC mediation team was the Constitution Amendment 18 which aimed to align the Zimbabwe election conduct with the SADC Principles and Guidelines Governing Democratic Elections. Other significant reforms instituted by the SADC facilitation team included exclusion of the police at polling stations, conducting vote counts at polling stations instead of transporting them to a central place, and posting of election results outside each station. To a larger degree, SADC leadership demonstrated their appreciation of the importance of peace in the conduct of elections. The relatively peaceful political atmosphere mediated by the Mbeki-led SADC facilitation team indeed, paid some dividends. Both the 29 March

2008 presidential and parliamentary, harmonised elections witnessed impressive civil participation (Raftopoulos, 2013). It is prudent to infer that the 29 March 2008 elections met the requirements for free and fair elections as stipulated by the SADC Principles and Guidelines on the conduct of elections. An interesting development is that for the first time since 1980, ZANU-PF lost the House of Assembly winning 97 seats, while the two MDC splinter groups combined polled 109 seats (Tsvangirai-led faction 99 and 10 seats for the Mutambara-led faction) (Raftopoulos, 2013; Chitiyo, 2009; Masunungure, 2011). This was a result of the encouraging political environment that allowed other parties to campaign freely.

The positive outlook of the entire election process was however spoiled by the Zimbabwe Election Commission (ZEC)'s decision to withhold the release of the presidential election outcome. This created anxiety and frustration among many Zimbabweans yearning for political change. The election results were only released after a month, on 2 May 2008, and this created pandemonium as the opposition and their supporters raised allegations of vote rigging (Raftopoulos, 2013; Bratton and Masunungure, 2008). The shocking outcome of the 2008 harmonised election was that ZANU-PF lost to the MDC-T by polling 43.2 per cent against the 47.9 per cent of the MDC-T (Masunungure, 2011). The technicality was however that, constitutionally, the MDC needed to win 51 per cent majority vote to be declared the outright-winner and as a result, a run-off election was to be conducted (Mapuva, 2010).

These developments require some incisive scrutiny of the operations of ZEC as an Election Management Body (EMB). From a governance view point, the decision to delay the release of the election outcome showed the ineptness of ZEC in handling electoral affairs of the country. Comparatively, unlike the Independent Electoral Commission (IEC) of South Africa, which is self-governing in its functions, ZEC operates purely from a government model. The Zimbabwe Electoral Act of 2004 talks about the establishment of an Independent Electoral Commission (in this case ZEC) responsible for administering the electoral affairs of the country. Interestingly, the Electoral Act (2004) further attests that the president is supreme over all the functions of ZEC, as he can appoint and fire commissioners. It is fair to say that because of such contradiction in terms, ZEC is constrained in terms of discharging its duties of managing the election processes of the country. Considering the partisan nature of ZEC where it is aligned to the president who is again, the head of ZANU-PF, one can safely claim that the EMB in Zimbabwe is captured by ZANU-PF. The delay in the release of the 2008 presidential election results was therefore no coincidence. As incisively noted by Mapuva (2010) since inception in 2004, ZEC as a public institution has not been democratised and political interference has for long negatively impacted on electoral integrity.

The fact that President Mugabe lost the election for the first time in his political career and that he had to find himself again pitted against Tsvangirai in a re-run, triggered panic and turmoil within the ZANU-PF political circles. It also meant that ZANU-PF had to brace itself again for a tight election contest and had to restrategise to save the face of the president. The next section therefore details the 2008 run-off election conduct and the robust SADC entry into the election conflict as well as the cross-border ramifications of the conflict.

Scope and nature of the 2008 run-off election violence in Zimbabwe

As alluded to earlier, the MDC's failure to claim an outright majority victory resulted in a run-off election, set for 27 June 2008. This time, ZANU-PF was forced to restrategise its election campaign to avoid another shame. As a result, state security agencies, namely the army, police and the intelligence supported by other non-statutory structures such as the war veterans and Border-Gezi youth militia, embarked on terror campaigns. The popular slogan in local parlance was that of Mugabe for 2008 and nothing more. The relatively conducive election environment that characterised the 29 March harmonised elections was replaced with a reign of terror (Masunungure, 2011). What worsened the situation was the Presidential Order, when the then President, Robert Mugabe urged ZANU-PF supporters to establish "an almost military/war like leadership which will deliver" (The Zimbabwe Independent, 2008: 1). Mugabe's declaration licenced wanton electoral related human rights abuses and the situation worsened when he openly declared that he would not give up power because of a mere X (The Zimbabwe Independent, 2008). The 2008 run-off election earned notoriety when some of the military chiefs declared their allegiance to ZANU-PF arguing that if ZANU-PF lost to the MDC, the army were ready to take over (Masunungure, 2011). These threats and the unspeakable levels of pre-electoral violence compelled the MDC-T Leader Morgen Tsvangirai to pull out of the electoral contest. Interestingly, President Mugabe audaciously embarked on a one-man election race and was declared a winner with 90, 22 per cent (ZESN, 2008). Mugabe's decision to run unopposed in the election attracted negative labels globally with scholars like Masunungure (2011) calling it "militarised election", while some sections of the media denigrated it as Africa's shame (*The Economist*, 2008).

The manifestation of election violence in Zimbabwe that involved state security agencies can be analysed from the party strength and authoritarian durability as discussed by Levitsky and Way (2010). These scholars propound that most ruling parties in authoritarian set-ups thrive politically by incorporating youth wings and other grassroots structures as part of the state's police power. Fundamentally, as explained by Levitsky and Way (2010), the touted distinction between the army and politicians has its roots in the liberation war where most of those who occupied key positions in the state security structures are former liberation fighters. Upon attaining independence in 1980, the leadership in Zimbabwe adopted the strategy of placing those with liberation credentials in key strategic ministries and institutions as a way of fostering cohesion between the politicians and the military. This observation helps to explain the army's loyalty and the authoritarian stability that prevailed for 37 years in Zimbabwe as well as the continuous youth factor in the commission of electoral violence in the country.

As if that was not enough, Tsvangirai's decision to pull out did not bring any stability. ZANU-PF embarked on another post-run-off witch hunt operation, code named Operation *Makavhoterapapi* ("whom did you vote for?"). Those who did not participate in the run-off election or who abstained from voting were also not spared by Operation *Makavhoterapapi* (Chitiyo, 2009). This operation saw harassment, beatings and torture of opposition party supporters. Electoral violence had a huge toll on ordinary citizens of Zimbabwe. Statistics show that over 250 people were killed as punishment for previously not voting for the ruling party while an estimated 36,000 people were internally displaced, with some crossing the borders into neighbouring countries (see detailed discussion below) for safety reasons (Human Rights Watch, 2011; Chitiyo, 2009). The involvement of state security clusters in the perpetration of electoral violence also meant that ZANU-PF party supporters and other agents of violence could enjoy impunity. This was due to selective application of the rule of law. Arguing from a more gendered perspective, Murithi and Mawadza (2011) lament that partial application of the rule of law exposed women to rape and other forms of gender based violence. In general, unspeakable electoral related human rights abuses revealed what the USAID (2010) calls the challenge of electoral security in Africa. In this case, it also meant that election stakeholders ranging from voters, the media, public officials and polling agents among others need protection during election times. Evidence of state orchestrated human rights violation prior to and after the June 2008 election run-off show that the ZANU-PF and President Robert Mugabe grossly undermined the cardinal principles underpinning democracy and good governance (Human Rights Watch, 2011; Howard-Hassmann, 2010). In fact, scaring the opposition out of the election contest is a typical example of coup by ballot. These developments triggered varying reactions by SADC member states. Equally important is to unpack the regional dimensions of the election conflict that also propelled prompt reaction by the regional leadership.

The cross-border dimensions of the electoral conflict in Zimbabwe

A perceptive submission by Carik (2009) is that cross-border human rights violations against civilians are endemic in African states at conflict or experiencing looming conflicts. This is again underscored by Willemse et al. (2015) who stress the need to understand the channels of connectivity and border governance to grasp cross-border dimensions of various conflicts. It is true that whenever there is a conflict in any given country, it comes with some associated massive displacement of people (Chingono and Nakana, 2009). Given the porous nature of borders and the transnational nature of intra-state conflicts in contemporary Africa, the safety and security of the displaced populations is compromised. More importantly, current typologies on migration in SADC do not reflect specifically on contemporary realities of electoral induced displacements. In the context of SADC, studies show that the 2008 run-off electoral conflict triggered massive movement of people, which had cross-border human security implications. Conspicuous in the Zimbabwe situation is what can be analysed as state or regime induced displacements. These include decline in confidence in the electoral processes in Zimbabwe, increasing unemployment rates and high incidence of electoral violence that caused massive displacements and migration of many Zimbabweans into neighbouring countries such as South Africa, Zambia and Botswana, among other relatively peaceful neighbouring states (Maphosa, 2010).

Conventionally, as noted by Chingono and Nakana (2009), the displacement of people and refugee crisis in the SADC has been attributed to wars of liberation. In post-Cold War Africa, intra-state conflicts in countries such as Somalia, Rwanda, Burundi, Mozambique and the DRC have their large share of human displacements throughout the region. The 2008 run-off election conflict in Zimbabwe however triggered a new wave of political and economic displacements with cross-border dimensions in the SADC region. Electoral related human displacements in Zimbabwe wreaked more havoc on Botswana and South Africa given their relatively stable economies by African standards. While millions were internally displaced by the outcome of the contested election, hyperinflation which peaked in 2008, created an unprecedented level of migration of thousands of Zimbabweans into neighbouring countries both legally and illegally in search of economic and political safety (Crush and Tevera, 2010; Hammar, McGregor and Landau, 2010). In the aftermath of the June 2008 run-off presidential elections in Zimbabwe, Botswana Foreign Minister Skelemani complained that Zimbabweans, who were fleeing the crisis in their own country, were draining his country's resources, and he appealed to the international community for assistance (The Zimbabwean, 2008). Statistics from the Botswana Immigration Office indicated an increasing number of Zimbabweans entering the country through formal border points, rising from 746,212 in 2006 to 1,041,465 in 2008 (Campbell and Crush, 2012). In the context of South Africa, relative economic and political stability as well as South Africa's hegemonic profile in the region renders it a regional magnet of migration. Consequently, by June 2008, an estimated 3.5 million Zimbabweans were reported to be living in South Africa either legally and illegally (Campbell and Crush, 2012). While some authorities give credence to economic and social benefits of migration, citing developments like remittances, skills sharing, employment creation and investments among others that can be tapped for the betterment of the host country (Boubtane and Dumont, 2013; Binci, 2012), the influx of economic and political refugees from Zimbabwe into South Africa presented the inverse.

Allegations of foreign nationals including Zimbabweans taking job opportunities for local youth, competition for services and amenities including housing and resulting xenophobia in host countries are well documented (Batisai, 2016; Crush and Ramachandran, 2014; Solomon and Kosaka, 2013). The spate of violence that engulfed most South African cities and semi-urban areas saw more than 60 people losing their lives while 1,000 more were displaced (Forced Migration Studies Programme – FMSP, 2009). Most of the Zimbabweans who were displaced by the election violence found themselves between the jaws of a pliers when they were confronted with the spate of hate and resentment in most densely populated areas of South Africa. Souring of relations were premised on allegations that most foreign nations (mainly Zimbabweans who constitute large numbers) were stealing jobs from locals and fraudulently getting access to houses provided under the Reconstruction and Development Programme (RDP), which was meant for previously disadvantaged South African citizens (Hadland, 2008). Reading from the verbal allegations and physical manifestations of hostilities towards Zimbabweans, one can argue that xenophobia as a form of collective violence against foreign nationals, Zimbabweans in this case, was perpetrated as a tool defining belonging, as well as for exclusion of the other.

From a regional integration point of view, one can argue that electoral displacements in Zimbabwe and their cross-border dimensions into South Africa triggered a xenophobic outrage from the host communities. This resonates with what Crush and Ramachandran (2014) describe as politics of belonging. The exclusionist rhetoric embedded in the expression of xenophobia in such derogatory labels as *Makwerekwere*, referring to non-South African nationals from other African countries, attest to the unstable social and political integration in the SADC region. Xenophobia in general threatened the very foundations of Ubuntu and regional cooperation that the founding architects of SADC endeavoured to nurture. It is also plausible to assert that xenophobic violence in the aftermath of state induced human displacements in Zimbabwe shows that the SADC as a region is integrated politically but not socially. In fact, at the grassroots level, SADC citizens are estranged from each other.

In addition to the cross-border dimensions of the crisis, the aspect of political spillover requires thorough examination. Perceptions that then South African President Thabo Mbeki was shielding Mugabe's despotism, which steered internal instability in the domestic sphere, also deserve further scrutiny. More importantly, in the aftermath of the 2008 election run-off, civil society and other pressure groups mobilised against Chinese arms shipment to Zimbabwe, at a time when the country was experiencing an unprecedented state repression. As argued by Du Plessis (2008), the Zimbabwean government neglected the human security needs of its people by prioritising arms procurement from China. The South African Trade and Allied Workers Union (SATAWU) instructed its workers not to offload the ship's cargo of arms, since they would be used against civilians. Together with the South African Litigation Centre (SALC), SATAWU challenged the morality of allowing transportation of weapons to a country already experiencing serious state-sponsored violence (Du Plessis, 2008). Such developments showed that the crisis in Zimbabwe had a potential to destabilise its neighbours. One would say that, by attempting to have the arms offloaded at the Durban port, South Africa's commitment to human rights promotion became exceedingly debatable. Many in South Africa started losing confidence in the ANC led government as it tried to propel tyrannical rule in their neighbourhood. In the domestic sphere of South Africa, there is a belief that Mbeki's handling of the 2008 electoral conflict in Zimbabwe contributed to his recall at the elective conference in Polokwane (Raftopoulos, 2008). Further evidence shows that South Africa was on the receiving end of the crisis in Zimbabwe, both socially and economically.

Gender dimensions of the pre- and post-2008 election violence in Zimbabwe

As introduced in the preceding discussions, one striking aspect of the 2008 electoral violence is that it impacted negatively on women and the defenceless of society. Women were subjected to untold suffering, from politically insti-gated violations to cross-border crimes in their desperate journeys. The pre- and post-2008 state orchestrated election violence in Zimbabwe defied any gender barrier. Women found themselves immersed in a social, economic and political conundrum. The outlawing of opposition politics as sell-outs forced many men in the rural settings to flee to cities or cross the borders for safety (Matanda, Rukondo and Matendera, 2016). As a result, relatives of the so called male "sell-outs", mainly females were subjected to torture and beatings especially in the rural areas. Harassment, torture and other gender-based violations were meant to force women to confess the whereabouts of their male relatives suspected of being MDC sympathisers. The 2014 report by the Research and Advocate Unit (RAU) revealed that 62 per cent of women sur-veyed in their study of the 2008 election violence reported to have experienced politically instigated violations (RAU, 2014).

Unprecedented levels of violence during the 2008, pre- and post-run-off election in Zimbabwe showed that the rule of law was transitorily suspended. Election related human rights abuses could be committed without any legal repercussions. Violation of women's rights was further aggravated by the military factor in the entire electoral conduct especially the Youth Militia in the villages. Reports show that women were forced to partake in political rallies especially during the so called *pungwes* or night rallies typical of the war of liberation struggle in the 1970s. This phenomenon was quite noticeable in the rural areas where attending rallies was declared compulsory (RAU, 2011). In some contexts, as exposed by Matanda, Rukondo and Matendera (2016) suspected MDC sympathisers were deprived of access to fertiliser, seed and other government rations meant to caution people during the height of the 2008 economic afflictions. It is apparent that violence was intended to instill fear and coerce women to vote for ZANU-PF. This was also meant to preserve ZANU-PF's supremacy in the rural settings. This scrutiny agrees with Honwana (2007)'s supposition that abuse of women during volatile political situations resonates with the quest for nourishing patriarchal and militaristic hegemony by many African countries. It can be further inferred that, while studies have exposed women's vulnerability during war situations (Bardall, 2011; Eisenstein, 2007; Ehrenreich, 2005), adoption of elections as an integral tool for democratisation has also deepened women's vulnerability due to election related violations. The reality is that elections in Zimbabwe have become a highly contested battle settled with bloodshed in most instances like the pre- and post-run-off environments in 2008. In view of ZANU-PF's waning popularity nationwide, resorting to violence served a dual purpose of suppressing civil dissent and manipulating the electoral system (Sachikonye, 2012). Since 2000, the opposition MDC has chastised the Zimbabwe electoral system for being flawed and partisan.

One other significant factor to note is that the 2008 harmonised presidential and parliamentary as well as the run-off elections were conducted at a time when Zimbabwe was in the doldrums in terms of economic performance. The country was undergoing politically induced economic distresses including unprecedented levels of inflation, health decline, acute food shortages, fuel crisis and high unemployment levels among other challenges. High unemployment rates and political repression forced most family men and able bodied young men to cross the frontiers into neighbouring countries (Matanda, Rukondo and Matendera, 2016; Duri et al., 2013) for safety and better opportunities. The result was that women had to devise some coping mechanisms in the face of economic meltdown and election violence. Likewise, as mentioned by Duri et al. (2013), cross-border trading emerged to be the dominant survival strategy resorted to by some women in both rural and urban areas. In the urban context, some men remained vigilant and pursued politics for the sake of change, while most women took the path of cross-border trading in order to fend for their families.

The process of cross-border scavenging by the desperate Zimbabwean women was characterised by some as serious human security challenges. Cross-border crime is one element of the human security implications of the electoral conflict that peaked after the 2008 election run-off. Interestingly, the Zimbabwe–South Africa border which is the hub of connectivity from Cape to Cairo, suddenly became the hub of criminal elements. The cost of securing a passport was quite high at the time and one had to part with 150 United States dollars to get a passport in Zimbabwe. As a result, most poor Zimbabweans who embarked on the desperate journey resorted to illegally crossing into South Africa (Reliefweb, 2010). These however encountered the criminal activities of thugs called *Magumaguma* around the Beitbridge border post. There are numerous reports of *Magumaguma* preying on illegal immigrants, especially women and girls, robbing them of their little possessions such as cell-phones, money and in some extreme situations raping and murdering them (International Organisation for Migration (IOM), 2013; Mahati, 2012). With no papers or legal documents to cross the border with, many women opted for illegal crossing into South Africa, and in the process some experienced incidence of rape. As noted by Mashiri (2013), many rape cases were not reported for fear of societal stigma and divorce by husbands. Literature reveals that *Magumaguma* employ the strategy of offering cross-border assistance to those without papers and in turn rob them or at times use violence directly, inflicting lasting damage on their victims (Maphosa, 2010; Raftopoulos and Mlambo, 2010). On the South African side, this had some defence and security implications as the government had to intensify border security for purposes of countering *Magumaguma*'s criminal activities in the thick forests along the Limpopo River. A human security study of the Zimbabwean migrants in South Africa conducted by Mawadza (2008) revealed elements of compromised safety and sanitation for many women. With no relatives or friends to cater for their accommodation needs, many were left with the option of living on the streets of major cities in South Africa. This phenomenon peaked during the height of the 2008 crisis and many women experienced violations on the streets.

In a nutshell, several dimensions of the crisis discussed so far, showed that as the people of Zimbabwe started crossing the frontiers of the country during the peak of the 2008 electoral conflict, the crisis assumed a regional complexion and ceased to be a Zimbabwe issue alone. This leads to the next section which focuses on how regional integration was preserved as countries tried to respond to the electoral conflict.

Analysis of the SADC reaction to the 2008 run-off election violence vis-à-vis regional integration

A cursory glance at the electoral conflict in Zimbabwe shows a tendency by authoritarian regimes to hijack democratic institutions for purposes of legitimacy. As aptly discussed by Levitsky and Way (2010), in the contemporary era there is a challenge of competitive authoritarians who perpetually participate in elections while constraining the playing field to favour the incumbents. The same challenge was faced by the Zimbabwe African People's Union (ZAPU) in the 1980s and 1990s (Levitsky and Way, 2010) where the playing field remained restrictive with the hope of preserving Zimbabwe as a *de facto* one party state. What can be deduced therefore from the election experience in Zimbabwe is a partial embracement of democratic precepts. By merely looking at the election driven violence in Zimbabwe, it is fair to assert that conducting elections is merely a democratic ritual. Way back, Sithole and Makumbe (1997: 134) attributed these developments to a "commandist political culture" imported from the liberation war conduct.

In retrospect, the outcome of the electoral conflict in Zimbabwe showed that the behaviour of the ZANU-PF led government of Zimbabwe characterised by the flagrant human rights violations as well as the contraventions of constitutional rule, contravened several regional and sub-regional statutes governing elections and democracy. These include the African Charter on

Human and Peoples' Rights, the AU's 2007 African Charter on Democracy, Elections and Governance and article 4 of the Declaration and Treaty of SADC. Fundamentally, article 4 of the Declaration and Treaty of SADC obligates member states to conduct themselves democratically by observing human rights as well as promoting peace and security in the region. The spillover into neighbouring countries and the contagion effects of the electoral crisis showed that the SADC community was faced with the mammoth task of holding a founding member of SADC (Ndlovu-Gatsheni, 2011) accountable for flagrantly disregarding the guidelines on elections.

Whatever decision they were to adopt finds legal basis in the above outlined statutes. Considering these incisive observations, the most unequivocal question is: what should have been done in view of flawed electoral processes characterised by abhorrent violation of human rights? Normatively, there is a global legitimate expectation that SADC would take robust intervention measures and hold the ZANU-PF led government of Zimbabwe accountable for the unprecedented and wanton electoral related human rights abuses in Zimbabwe. Again, the efficacy of SADC as a regional institution was put to test. Since SADC is an intergovernmental organisation (IGO) involving state parties as principal actors, its reaction should also be interpreted in line with the SADC precepts regarding peaceful resolution of disputes. It is important to stress that electoral conflict in Zimbabwe presented the SADC bloc with a dual challenge of upholding the fundamental rights and freedoms of the people of Zimbabwe and at the same time ensuring regime security, regional integration and cooperation. Varying reactions by member states show that the 2008 run-off electoral contest in Zimbabwe nearly fractured SADC integration. The decision by then President Robert Mugabe to conduct a one-man election race and disregard human rights as evidenced by the unprecedented levels of pre- and post-run-off electoral violence attracted wide condemnation from concerned SADC member states. Zambia and Botswana were the loudest critics of the political developments in Zimbabwe. Unlike the then President Thabo Mbeki who was the chief mediator and maintained a mild approach to the political developments in Zimbabwe by opting for "Quiet Diplomacy", Botswana through its foreign affairs Minister Phendu Skelemani condemned electoral related human rights abuses in Zimbabwe.

More importantly, the no crisis proclamation by Mbeki did not go down well with the leadership in Botswana, who interpreted it as a demonstration of indifference to the electoral conflict in Zimbabwe (Rossouw and Moyo, 2008). The rift among the SADC members widened when President Ian Khama rebuked Mugabe for holding on to power illegally. Fundamentally, Botswana condemned ZANU-PF for not observing both the AU and SADC guidelines governing the conduct of democratic elections. As if that was not enough, President Khama lobbied for delegitimising Mugabe by not recognising his presidency and suspending Zimbabwe from SADC (Open Society Initiative for Southern Africa, 2013; Piet, 2009; Diakanyo, 2008). To bolster his position, President Khama boycotted a number of SADC meetings in the aftermath of the run-off elections and went as far as threatening closure of the High Commission of Zimbabwe in Botswana (Open Society Initiative for Southern Africa, 2013; Diakanyo 2008). While the government of Zimbabwe retaliated by accusing Botswana of supporting opposition politics and interfering in Zimbabwe's internal affairs, Botswana's lamentations were a clearcut pointer to the impending disintegration of SADC.

While Botswana remained resolute in calling for a democratic atmosphere in Zimbabwe and condemning the ZANU-PF led government for electoral related human rights abuses, it had an abrupt change of opinion by demanding that the West revoke sanctions imposed on Zimbabwe to allow the situation in the country to redressed (Piet, 2010). These remarks made in 2010 and the sudden change in attitude by Botswana speaks to the power of regional integration in SADC. Fears of being labelled a Western puppet and being isolated for deviating from the SADC precepts of non-confrontation might have informed Khama's later position. As alluded to earlier, Botswana was not a solo voice in raising concern over deteriorating electoral conditions in Zimbabwe. The then Zambian President and SADC Chairperson, Levy Mwanawasa vehemently criticised SADC for its big brother approach towards Mugabe.

More importantly, during the 12 April 2008 extra-ordinary Summit of Heads of State and Government in Lusaka, Zambia, Mwanawasa openly criticised SADC's insistence on quiet diplomacy instead of taking robust and practical action on the electoral conflict in Zimbabwe (SADC Communiqué, 2008). As a way towards watering down the simmering tensions in SADC over the electoral conflict in Zimbabwe, Thabo Mbeki, the SADC chief mediator, shocked the entire world when he proclaimed that there was no crisis in Zimbabwe (Mail&Guardian, 2008). One can safely say that with such carefree pronouncements, Mbeki flouted the wishes of the Zimbabweans yearning for regime change. In addition, incongruences in interpreting the electoral crisis in Zimbabwe showed how the electoral conflict in Zimbabwe pitted the SADC member states against each other. It is evident from the divergences in opinion that the electoral conflict was new to the region, hence there existed a lack of expertise in dealing with such. As alluded to, SADC was caught between protecting human rights and preserving SADC regional integration. The death of the Zambian President Levy Mwanawasa on 19 August 2008, gave Mbeki and his mediation team a window of opportunity for invoking quiet diplomacy as well as adopting a state-centric approach to resolving the electoral conflict.

Though conscious of human rights abuses and the deteriorating human security situation in the post-2008 run-off era, SADC capitalised on divergences in crisis naming and interpretation. The strategy of quiet diplomacy with a preventive element was invoked in order to resolve the election impasse. This involved constructive engagement with the conflicting parties, namely, ZANU-PF and the MDC (Landsberg, 2016; Murithi and Mawadza, 2011). Mbeki took his mediation leverage to encourage dialogue and

negotiation in line with the wisdom of African solutions to African problems. The preferred approach also resonated with the SADC dictum on
peaceful resolution of disputes as enshrined in the Declaration and Treaty
of SADC. From a normative reasoning, such developments concur with the
understanding of institutional theory which stipulates that political behaviour either individually or as a collective is shaped by norms informed by
membership of the institutions. These norms define the logic of appro-
priateness (March and Olsen, 1996).

As part of the broader plan to save regional integration, Mbeki managed to
defuse the electoral conflict by invoking the liberation war rhetoric. The
foundation upon which SADC was established involves memories of the liberation war, from the Front-Line States, SADDC and down to the current
SADC. On another angle, Mbeki managed to convince the regional leadership on his take on the crisis, especially that Mugabe was in fact a victim in
the whole election debacle. As such, the forces of neo-colonialism were at
work trying to punish Mugabe for his astute anti-colonial position (Phimister
and Raftopoulos, 2004). It also implies that when Mbeki made the "no crisis"
pronouncement, it was a smooth reminder that SADC should not compro-
mise its founding institutional principles in its handling of the electoral conflict in Zimbabwe. One can therefore say regional integration was of more
concern than the deteriorating human security conditions because of the
election conflict.

From another dimension, the MDC's lack of revolutionary foundation
presented it as a misfit to SADC regional cooperation. More so, the vilifica-
tion of MDC as a Western puppet appealed most to the SADC leadership
and it explains SADC's empathy towards ZANU-PF. It would seem like a
resolution to the conflict that disfavoured ZANU-PF would mean a reversal
of the gains of liberation struggle. This is a path that SADC was not prepared
to follow given that President Mugabe is revered as a liberation cult hero
regionally. More so, the geopolitical configurations of the SADC region are
another metric that helps to explain SADC's state model approach to resolving the 2008 run-off election conflict. The SADC decision to adopt a political solution that provided ZANU-PF with a soft landing is not isolated. It is
important therefore to say that historical factors played a significant role in
the establishment of SADC. These factors are also referenced as crucial in the
resolution of conflicts in the region. A more nuanced analysis by Levitsky and
Way (2010) shows that non-material means of cohesion helped to sustain
SADC integration in view of the fierce election battle. Factors such as ideology, shared history and memories of the liberation struggle were deemed vital
for regional solidarity. The argument is that countries with a shared colonial
experience and revolutionary governments usually overcome opposition or
any international pressure by merely invoking their solidarity embedded in
ideology. This may explain why SADC managed to handle the electoral crisis
in Zimbabwe without suffering what Levitsky and Way (2010) describe as
debilitating defections.

Having integration cemented by the irrevocable liberation history as well as shared colonial experiences, suggests that the political imperatives of the region could not be twisted for any other reason. One can therefore argue that revolutionary regimes are still a political force to reckon with, in the SADC. South Africa is governed by the African National Congress (ANC), Namibia by South West People's Organisation (SWAPO), Chama cha Mapinduzi (CCM) in Tanzania, Popular Movement for the Liberation Front of Angola (MPLA) and the Mozambican Liberation Front (FRELIMO). This aptly suggests that the whims of liberation struggle and not election guidelines matter most as far as SADC integration is concerned. More importantly, South Africa though now a regional hegemony, owes its political liberation to the support provided by its fellow SADC comrades, Zimbabwe included (Hulse, 2016; Adolfo, 2009; Flemes, 2009). The historical part that informed the resolution of the electoral conflict showed that South Africa, as the SADC chief mediator, was not prepared to concede to Western demands for isolating Mugabe. A critical reflection on the history of SADC formation provided by Flemes (2009) shows that the doctrine of counter hegemony premised on discouraging apartheid dominance of the region was crucial. Coming from a long history of apartheid's destructive engagement, it would seem like South Africa did not want to be projected along that direction, hence the decision to lead SADC towards a constructive engagement path towards resolving the election tension in Zimbabwe.

Instead, guided by the state-centric approaches and what Adolfo (2009: 7) calls the "politics of solidarity", the institutionalisation of Government of National Unity (GNU) was deemed an alternative path towards resolving the electoral conflict. This followed the 28th Ordinary SADC Summit convened in Sandton, South Africa between 16 and 17 August 2008 (SADC Communiqué, 2008). The decision to resolve the electoral conflict through unity government resulted in the signing of the Global Political Agreement between ZANU-PF and the two MDC formations (One faction led by Morgen Tsvangirai and the other led by Arthur Mutambara) on 15 September 2008. To show that integration and cooperation mattered most for SADC, on 11 February 2009 during the inauguration of the GNU, Mugabe was appointed President, while Morgan Tsvangirai assumed the position of Prime Minister and Mutambara was appointed Deputy Prime Minister (MoU, 2008). One can argue that the political path that was adopted in resolving the 2008 run-off election conflict in Zimbabwe falls short of measures that promote democracy, human rights and economic development of the country. Again, the GNU solution has been critiqued for not being people orientated and lacking sustainability concerning the future of elections in the country (Mapuva, 2010).

Fundamentally, the non-confrontational approach adopted by the Mbeki led mediation team, helped to sustain SADC relations before the election conflict in Zimbabwe. Unlike the calls by the West for confrontational approaches, the chapter reveals that Mbeki and his team followed the same trajectory of preventive diplomacy characterised by dialogue and persuasion. This state centred approached helped to glue SADC countries together. Again, although not a popular solution, it emerged that SADC managed to preserve its regional stability, integration and cooperation by institutionalising the GNU in 2009. As captured in the chapter discussions, SADC also managed to sustain its integration during the electoral conflict in Zimbabwe by adhering to the accepted norm on peaceful settlement of disputes. In a nutshell, despite some divergences in opinions on the electoral conflict in Zimbabwe, the founding principles of SADC embedded in the liberation ties are integral to regional integration and cooperation. This also explains Botswana's change of opinion in interpreting the electoral conflict in Zimbabwe.

Conclusion

The prime focus of this chapter was to examine how the highly militarised 2008 run-off electoral conflict in Zimbabwe impacted on SADC regional integration and cooperation. The chapter revealed that SADC failed to handle the electoral situation in Zimbabwe, which led to large scale movements of people from Zimbabwe to many SADC countries including South Africa. This shows that at the level of the ordinary person, SADC failed to achieve peace. Even though the SADC states colluded to protect a dictator in the name of regional integration, the massive migration of Zimbabweans to neighbouring countries actually destabilised the region contributing to xenophobic attacks in Botswana and South Africa. All this can be traced to the SADC handling of the election run-off. As demonstrated by the study, it is fair to argue that SADC states wanted to protect a dictator for the sake of championing fake Afro-radicalism and the defence of independence (Ndlovu-Gatsheni, 2006) against the so called vampirism of neo-colonial matrices of power (Ndlovu-Gatsheni, 2013). Further examination of the human security costs of the electoral conflict points to the fact that regional integration (defined as the club of people/leaders who fought against colonialism) should exist at the expense of the peace of the ordinary person, as long as the SADC has prevailed against the Western threat – real or imagined.

Reading from the SADC reaction to the election conflict in Zimbabwe, one can argue that the challenge posed by the MDC during the 2008 elections in fact gave the regional leadership an opportunity to rebuild their organisation. The regional body was actually afforded an opportunity for renewal by showing collective solidarity against a non-revolutionary party. Though initially vocal, the sudden turn by President Ian Khama of Botswana to a greater degree exposed the regional costs of defection as he ended up being a solo voice against the Mugabe regime. What one can argue is that SADC countries managed to shelve their differences in the face of opposition adversity and preserved their identity as a revolutionary organisation.

A critical reflection on the cross-border dimensions of the electoral conflict showed that the Zimbabwean regime induced displacements which caused ethnic tensions in the receiving countries and this further strengthened the argument that elections are increasingly becoming a destabilising force to the region and with the potential to rupture regional peace, security and integration. As revealed in the chapter, the predicament that SADC found itself in is largely a result of lack of an electoral security framework and this makes it imperative for regional leadership to craft and institute policies for the purposes of mitigating election related conflicts.

A more worrying development is that of the military takeover in Zimbabwe which again paints a gloomy picture of the future of elections in the country. The fact that the state security apparatus has been involved in the commission of and command responsibility in election related human rights abuses may be an indicator of the high probability of recurrence of election-driven violence or conflicts. This speaks therefore to need for harmonising election security policies with the regional integration statutes or else ordinary people in the SADC will continue to suffer violence, human rights abuse, displacement, insecurity and death, as the case of the Zimbabwean elections show.

References

Adolfo, E. 2009. *The Collision of Liberation and Post-Liberation Politics within SADC: A Study on SADCand the Zimbabwe Crisis.* FOI, Swedish Defense Research Agency. http://foi.se/ ReportFiles/foir_2770.pdf (accessed 13 October 2017).

Bardall, G. 2011. Breaking the Mold: Understanding Gender and Election Violence. International Federation for Electoral Systems, Washington, DC. www.ifes.org/p ublications/breaking-mold-understanding-gender-and-electoral-violence (accessed 5 November 2018).

Barnekow, S. E., and Kulkarni, K. G. 2017. Why Regionalism? A Look at the Costs and Benefits of Regional Trade Agreements in Africa. *Global Business Review* 18(1): 99–117.

Batisai, K. 2016. Interrogating Questions of National Belonging, Difference and Xenophobia in South Africa. *Agenda* 30(2): 119–130.

Bekoe, D. 2010. Trends in Electoral Violence in Sub-Saharan Africa. Peace Brief 13, United States Institute of Peace. www.usip.org (accessed 23 November 2017).

Binci, M. 2012. The Benefits of Migration. *Economic Affairs* 32(1): 4–9.

Boubtane, E., and Dumont, J. C. 2013. *Immigration and Economic Growth in the OECD Countries 1986–2006: A Panel Data Analysis.* Documents de Travail du Centre d'Economie de la Sorbonne, No. 2013.3. http://ftp://mse.univparis1.fr/pub/m se/ CE (accessed 12 October 2017).

Bratton, M., and Masunungure, E. 2008. Zimbabwe's Long Agony. *Journal of Democracy* 19(4): 41–55.

British Broadcasting Corporation (BBC). 2008. Botswana Urges Region Not to Recognize Mugabe's Re-election, 4 July.

Brown, M. M. 2003. Democratic Governance: Toward a Framework for Sustainable Peace. *Global Governance* 9(2): 141–146.

Campbell, E., and Crush, J. 2012. Unfriendly Neighbours: Contemporary Migration from Zimbabwe to Botswana. Migration Policy Series, 61. www.africaportal.org/p ublications/unfriendly-neighbours-contemporary-migration-from-zimbabwe-to-botswana/ (accessed 4 March 2019).

Carik, D. S. 2009. Porous Borders and the Insecurity of Civilians: Cross-Border Violence in Darfur and Eastern Chad. Ford Institute for Human Security Policy Brief.
www.fordinstitute.pitt.edu/Portals/0/General_PDF/Porous%20Borders%
20and%20the%20Insecurity%20of%20Civilians_ Carik.pdf (accessed 16 October 2017).

Chingono, M., and Nakana, S. 2009. The Challenges of Regional Integration in
Southern Africa. *African Journal of Political Science and International Relations* 3(10),
396–408.

Chitiyo, K. 2009. *The Case for Security Sector Reform in Zimbabwe.* Occasional
Paper, September. London: Royal United Services Institute.

Collier, P., and Vicente, C. V. 2012. Violence, Bribery, and Fraud: The Political Economy of Elections in sub-Saharan Africa. *Public Choice* 153(1–2): 117–147.

Crush, J., and Tevera, D. S. (Eds). 2010. *Zimbabwe's Exodus: Crisis, Migration, and
Survival.* Ottawa, ON: SAMP in cooperation with IDRC.

Crush, J., and Ramachandran, S. 2014. Xenophobic Violence in South Africa:
Denials, Minimalism, Realism, Casual Factors and Implications. *Migration Policy
Series* 66: 1–44.

Diakanyo, S. 2008. The Nonsensical Boycott of the SADC Meeting by Botswana. *Mail&
Guardian*, 16 August.http://thoughtleader.co.za/sentletsediakanyo/2008/08/16/the-
nonsensical-boycott-of-the-sadc-meeting-by-botswana/ (accessed 20 February 2018).

Du Plessis, M. 2008. Chinese Arms Destined for Zimbabwe over South African Territory:
The R2P Norm and the Role of Civil Society. *African Security Review* 17(4): 17–29.

Duri, K., Stray-Pedersen, B., and Muller, F. 2013. HIV/AIDS: The Zimbabwean
Situation and Trends. *American Journal of Medical Research* 1(1): 15–22.

The Economist. 2002. News Report: Mugabe's Smash-and-Grab – A Coup by Any
Other Name, 14 March. www.economist.com/node/1034280 (accessed 2 July 2017).

The Economist. 2008. Zimbabwe: Africa's Shame, 17 April.www.economist.com /node/
11052889 (accessed 20 October 2017).

Ehrenreich, B. 2005. *Abu Ghraib: The Politics of Torture.* Berkeley: North Atlantic Books.

Eisenstein, Z. 2007. *Sexual Decoys: Gender, Race and War in Imperial Democracy.*
New York: Palgrave.

Fischer, J. 2002. Electoral Conflict and Violence: A Strategy for Study and Prevention.
IFES White Paper, 1. http://aceproject.org/ero-en/topics/elections-security/
UNPAN019255.pdf/view (accessed 15 November 2013).

Fisher, J. 2013. *Elections and Conflict in Sub-Saharan Africa 2013: Somali Land, Cote
D' ivoire and Kenya.* Princeton, NJ: Woodrow Wilson School of Public and International Affairs, Graduate Policy Workshop.

Flemes, D. 2009. Regional Power South Africa: Co-operative Hegemony Constrained
by Historical Legacy. *Journal of Contemporary African Studies* 27(2): 135–157.

Forced Migration Studies Programme. 2009. Database on Responses to May 2008
Xenophobic Attacks in South Africa. 1(32): 104, 157.

Gerenge, R. 2015. Preventive Diplomacy and the AU Panel of the Wise in Africa's
Electoral-related Conflicts. South Africa Institute of International Affairs (SAIIA)
Policy Briefing 136.

Goldsmith, A. A. 2015. Electoral Violence in Africa Revisited. *Terrorism and Political
Violence* 27(5): 818–883.

Guichaoua, A. 2017. Elections in Africa: Democratic Rituals Matter Even Though
the Outlook is Bleak. *The Conversation.* http://theconversation.com/elections-in-a
frica-democratic-rituals-matter-even-though-the-outlook-is-bleak-85617?utm_m
edium=email&utm_ campaign (accessed 17 October 2017).

Hadland, A. (Ed.). 2008. Violence and Xenophobia in South Africa: Developing Consensus, Moving to Action. A partnership between the Human Sciences Research Council (HSRC) and the High Commission of the United Kingdom, October. http://ecommons.hsrc.ac.za/handle/20.500.11910/5188 (accessed 19 November 2017).

Hammar, A., McGregor, J., and Landau, L. 2010. Introduction. Displacing Zimbabwe: Crisis and Construction in Southern Africa. *Journal of Southern African Studies* 36(2): 263–283.

Honwana, J. 2007. *Child Soldiers in Africa.* Philadelphia: University of Pennsylvania Press.

Howard-Hassmann, R. 2010. Mugabe's Zimbabwe, 2000–2009: Massive Human Rights Violations and the Failure to Protect. *Human Rights Quarterly* 32(4): 898–920.

Hulse, M. 2016. Regional Powers and Leadership in Regional Institutions: Nigeria in ECOWAS and South Africa in SADC. KFG Working Paper Number 76, Berlin. www.polsoz.fu-berlin.de/en/v/transformeurope/publications/working_paper/wp/ wp 76/WP_76_ Hulse_ PRINT.pdf (accessed 23 November 2017)

Human Rights Watch. 2011. Zimbabwe: No Justice for Rampant Killings, Torture; Impunity Fuels New Abuses, Imperils Future Election. www. hrw.org/news/ 2011/ 03/08/zimbabwe-no-justice-rampant-killings-torture (accessed 23 November 2017).

International Organisation for Migration (IOM). 2013. Forum Responds to Cross Border Migration Challenges between South Africa and Zimbabwe. https://southa frica.iom.Int/news/forum-responds-cross-border-migration-challenges-between-south-africa-and-zimbabwe (accessed 23 November, 2017).

Landsberg, C. 2010. Pax-South Africana and the Responsibility to Protect. *Global Responsibility to Protect* 2(4): 436–457.

Landsberg, C. 2013. The Southern African Development Community's Decision-making Architecture. In C. Saunders, G. A. Dzinesa and D. Nagar (Eds). *Region-Building in Southern Africa: Progress, Problems and Prospects.* London: Zed Books, pp. 67–77.

Landsberg, C. 2016. African Solutions for African Problems: Quiet Diplomacy and South Africa's Diplomatic Strategy Towards Zimbabwe. *Journal for Contemporary History* 41(1): 126–148.

Levitsky, S., and Way, L. A. 2010. Beyond Patronage: Ruling Party Cohesion and Authoritarian Stability. Paper prepared for the American Political Science Association (APSA) Annual Meeting, Washington DC. https://ssrn.com/abstract=1643146 (accessed 12 November, 2017).

Lindberg, I. S. 2009. The Power of Elections Revisited. In I. S. Lindberg (Ed). *Democratisation by Elections: A New Mode of Transitions.* Baltimore, MD: John Hopkins University Press, pp. 314–341.

Mahati, S. T. 2012. Children Learning Life Skills through Work: Evidence from the Lives of Unaccompanied Migrant Children in a South African Border Town. In G. Spittler and M. Bourdillon (Eds), *African Children at Work: Working and Learning in Growing Up for Life.* Berlin: LIT Verlag, pp. 249–278.

Mail&Guardian. 2008. No Crisis in Zimbabwe, Says Mbeki, 12 April. https://mg.co.za/a rticle/2008–2004–12-no-crisis-in-zimbabwe-says-mbeki (accessed 16 September 2017).

Makumbe, J. M. W. 2002. Zimbabwe's Hijacked Election. *Journal of Democracy* 13(4): 87–101.

Maphosa, F. 2010. Transnationalism and Undocumented Migration between Rural Zimbabwe and South Africa. In J. Crush and D. Tevera (Eds), *Zimbabwe's Exodus:*

Crisis, Migration and Survival. Ottawa, ON: SAMP in cooperation with IDRC, pp. 345–362.

Mapuva, J. 2010. Government of National Unity (GNU) as a Conflict Prevention Strategy: Case of Zimbabwe and Kenya. *Journal of Sustainable Development in Africa* 12(6): 247–263.

Mapuva, J., and Muyengwa-Mapuva, L. 2014. The SADC Regional Bloc: What Challenges and Prospects for Regional Integration? *Journal of Law, Democracy and Development* 18: 22–36.

March, J. G., and Olsen, J. P. 1996. *Democratic Governance.* New York: Free Press.

Mashiri, L. 2013. Conceptualisation of Gender Based Violence in Zimbabwe. *International Journal of Humanities and Social Science* 3(15): 94–103.

Masunungure, E. V. 2011. Zimbabwe's Militarized, Electoral Authoritarianism. *Journal of International Affairs* 65(1): 47–64.

Matanda, D., Rukondo, H., and Matendera, E. 2016. Women and State Violence in Zimbabwe, 2000–2008. In J. Etin (Ed.), *Introduction to Gender Studies in East and Southern Africa.* Boston, MA: Sense Publishers, pp. 257–275.

Mawadza, A. 2008. The Nexus Between Migration and Human Security: Zimbabwean Migrants in South Africa. Institute of Security Studies Paper 162, Pretoria.

Memorandum of Understanding (MoU). 2008. Memorandum of Understanding between the Zimbabwe African National Union-Patriotic Front and the Two Movements for Democratic Change Formations, Harare, 15 September.www.peaceau.org/up loads/zimbabwe-memorandum-of-understanding.pdf (accessed 20 July 2017).

Mhandara, L., and Pooe, A. 2013. Mediating a Convoluted Conflict: South Africa's Approach to the Inter-party Negotiations in Zimbabwe. The African Centre for the Constructive Resolution of Disputes (ACCORD) Occasional Paper Series: Issue (1), 1–38.

Mlambo, A. S., and Raftopoulos, B. 2010. The Regional Dimensions of Zimbabwe's Multi-Layered Crisis: An Analysis. Conference Paper presented on the theme, "Election Process, Liberation Movements and Democratic Change in Africa", Maputo (supported by the Ministry of Foreign Affairs, Norway), 18 April. www. iese.ac.mz/lib/publication/proelit/Alois_Mlambo.pdf (accessed 16 August 2017).

Motsamai, D. 2010. When Elections Become a Curse: Redressing Electoral Violence in Africa. Electoral Institute of Southern Africa (EISA), Policy Brief Series (1), 1–17.

Murithi, T., and Mawadza, A. 2011. Voices from Pan-African Society on Zimbabwe: South Africa, the African Union and SADC. In T. Murithi and A. Mawadza (Eds) *Zimbabwe in Transition: A View from Within.* Johannesburg: Jacana Media, pp. 290–311.

Muzondidya, J. 2009. From Buoyancy to Crisis: 1980 to 1997. In B. Raftopoulos and A. S. Mlambo (Eds), *Becoming Zimbabwe: A History from the Precolonial Period to 2008.* Harare: Weaver Press, pp. 167–200.

Ndlovu-Gatsheni, S. J. 2006. *The Nativist Revolution and Development Conundrums in Zimbabwe* Occasional Paper Series: Volume 1, Number 4. Durban: ACCORD.

Ndlovu-Gatsheni, S. J. 2011. Reconstructing the Implications of Liberation Struggle History on SADC Mediation in Zimbabwe. South Africa Institute of International Affairs, Occasional Paper (92), 1–23.

Ndlovu-Gatsheni, S. J. 2013. *Coloniality of Power in Postcolonial Africa: Myths of Decolonization.* Dakar: CODESRIA.

Ngoma, N. 2003. SADC: Towards a Security Community? *African Security Review* 12 (3): 17–28.

Open Society Initiative for Southern Africa (OSISA). 2013. *Khama's Megaphone Diplomacy.* www.osisa.org/hrdb/blog/khamas-megaphone-diplomacy (accessed 20 February 2018).

Phimister, I., and Raftopoulos, B. 2004. Mugabe, Mbeki and the Politics of Anti-Imperialism. *Review of African Political Economy* 101(31): 385–400.

Piet, B. 2009. Khama, Mugabe Face-Off in the Offing at SADC Summit. *MmegiOnline,* 23 January. http://allafrica.com/stories/200901260241.html (accessed 12 October 2017).

Piet, B. 2010. Botswana Warms up to Zimbabwe. *Mmegionline,* 15 October. www.mmegi. bw/index.php?sid=1&aid=5680&dir=2010/ October/Friday15/ (Accessed 11 June 2011).

Raftopoulos, B. 2008. Reshaping Politics through Displacements. Key Note Address to the Conference on Political Economy of Displacement in Post-2000 Zimbabwe, Johannesburg, 9–11 June. www.tandfonline.com/doi/pdf/10.1080/ 03057070.2010. 485779 (accessed 16 November 2017).

Raftopoulos, B. 2013. An Overview of the GPA: National Unity. In B. Raftopoulos (Ed.). *The Hard Road to Reform: The Politics of Zimbabwe's Global Political Agreement.* Harare: Weaver Press, pp. 1–38.

Raftopoulos, B., and Mlambo, A. S. 2010. *Becoming Zimbabwe: A History of Zimbabwe from the Pre-colonial Period to 2008.* Johannesburg: Jacana Press.

Reliefweb 2010. 300 Zimbabweans Arriving in SA Daily: MSF. 14 May. Available at https://reliefweb.int/report/zimbabwe/300-zimbabweans-arriving-sa-daily-msf (Accessed 25 February 2019).

Research and Advocacy Unity (RAU). 2011. Women and Political Violence: An Update. Report Prepared by the Women's Programme Research and Advocacy Unity. http://researchandadvocacyunit.org/publication/gender-zimbabwe-5 (accessed 1 November 2018).

Research and Advocacy Unity (RAU). 2014. Women and Elections in Zimbabwe: Insights from the Afro Barometer. www.salo.org.za/wp-content/uploads/2014/08/ Women-and-Elections-in-Zimbabwe-Insights-from-the-Afrobarometer.pdf (accessed 28 October 2019).

Rossouw, M., and Moyo, J. 2008. Botswana Raps "No Crisis" Mbeki. *Mail&Guardian,* 8 April.http://mg.co.za/article/2008-04-18-botswana-raps-no-crisis-mbeki (accessed 11 November 2017).

Sachikonye, L. 2012. *Zimbabwe's Lost Decade: Politics, Development and Society.* Harare: Weaver Press.

Schoeman, M. 2009. From SADCC to SADC and Beyond: The Politics of Economic Integration. www.alternative-regionalisms.org/wp-content/uploads/2009/07/schoema r_fromsadcctosadc.pdf (accessed 27 November 2017).

Sithole, M., and Makumbe, J. 1997. Elections in Zimbabwe: The ZANU (PF) Hegemony and Its Incipient Decline. *African Journal of Political Science* 2(1): 122–139.

Solomon, H. and Kosaka, H. 2013. Xenophobia in South Africa: Reflections, Narratives and Recommendations. *Southern African Peace and Security Studies* 2(2): 5–30.

Southern African Development Community (SADC). 1992. Declaration and Treaty of SADC. www.sadc.int/documents-publications/show /Declaration__Treaty_of_ SADC.pdf (accessed 11 March 2018).

Southern African Development Community (SADC). 2000. *Communiqué of the SADC Summit of Heads of State and Government on Zimbabwe, 6–7 August, Windhoek-Namibia.* www.sadc.int/files/3913/5292/8384/SADC_SUMMIT_COM MUNIQUES_1980–2006.pdf (accessed 12 July 2017).

Southern African Development Community (SADC). 2008. *Communiqué of the SADC Extraordinary Summit on the Situation in Zimbabwe, Dar es Salaam-Tanzania, 28–29 March*. www.sadc.int/news/news_details.php?news_id=927 (accessed 23 May 2017).

Straus, S., and Taylor, C. 2009. Democratisation and Electoral Violence in Sub-Saharan Africa, 1990–2007. Paper delivered at the American Political Science Association, 3–6 September.

Taylor, I. 2014. Community of Insecurity: SADC's struggle for Peace and Security in Southern Africa. *The Round Table: The Commonwealth Journal of International Affairs* 103(1): 135–137.

United States Agency for International Development (USAID). 2010. *Electoral Security Framework: Technical Guidance Handbook for Democracy and Governance Officers*. Washington, DC: Creative Associates International. Wallsworth, G. 2015. Electoral Violence: Comparing Theory and Reality. http://econ.msu.edu/seminars/docs/WallsworthElectionViolenceSept2014Draft.pdf (accessed 22 December 2017).

Wendt, A. 1995. Constructing International Politics. *International Security* 20(1): 71–81.

Willemse, K., de Bruijn, M., van Dijk, H., Both, J., and Muiderman, K. 2015. What are the Connections between Africa's Contemporary Conflicts? *The Broker*. http://TheBrokerOnline_What-are-the-connections-between-Africas-contemporary-conflicts.pdf (accessed 12 September 2017)

The Zimbabwean. 2011. Army General, Nyikayaramba Vows not to Salute Tsvangirai. www.thezimbabwean.co/2011/05/army-general-nyikayaramba-vows-not-to-salute-tsvangirai/ (accessed 25 November 2017).

Zimbabwe Electoral Act. 2004. www.parliament.am/library/norelectoral%20law/zimbabve.pdf (accessed 11 March 2018).

Zimbabwe Election Support Net Work (ZESN). 2008. Report on the Zimbabwe 29 March 2008 Harmonized Elections and 27 June Presidential Run-Off. https://akcampaign.files.wordpress.com/2012/02/report-on-the-zimbabwe-29-march-2008-harmonized-elections-and-27-june-presidential-run-off.pdf (accessed 18 November 2017).

The Zimbabwe Independent. 2008. News Report: Mugabe Orders "Warlike" Campaign, 23 May.

10 Does war/conflict hinder REI in Africa?

The regional integration–conflict/war nexus

Ntongwa Bundala

Introduction

The objectives of the African Union (AU) are to bring about political, economic and social integration between member countries and make the continent a better place. It is only in such a way that Africa can achieve the desired ends, as highlighted by the Abuja Treaty, 1991 (African Union, 2005). The AU faces many challenges of achieving its objectives, one of which is the failure of Regional Economic Integration Plans (REIPs) in Africa. This threatens the achievement of the fourth aspiration of the people of Africa of building a peaceful and secure Africa as per Agenda 2063 (AUC, 2015). Agenda 2063 states that regional integration in Africa is a precondition for a peaceful and conflict-free continent (AUC, 2015). Agenda 2063 plans to, among other things, "silence the guns" and achieve peace in African countries by 2020 (AUC, 2015). It assumes positive impacts of regional integration on war and the prevention of conflict. The Agenda emphasises that regional integration is a precondition for peace and a secure continent. This means the achievement of Agenda 2063 highly depends on the success of regionalism. In other words, war and conflict in Africa might be a sign that regional integration has failed on the continent.

However, the debate on whether war and conflict influence/hinder the success of regionalism in Africa is ongoing. In this debate, the burden of the failure of regional integration seems to be totally shifted towards war and conflict as the major cause (Bah, 2013; Bah and Tapsoba, 2010; Ikubaje and Matlosa, 2016; Healy, 2011). Some studies support the assumption of Agenda 2063 that war and conflict hinder integration. But, how valid is the assumption? Analysis of this assumption is necessary under the leading question, "does the absence of war guarantee peace that ensures distribution, location and spatial organisation of economic activities, leading to successful regional integration?" In order to address this question, this chapter engages in a thematic analysis of the regional integration–conflict war–peace nexus. The chapter investigates whether the absence of war guarantees peace that ensures that the distribution, location and spatial organisation of economic activities leads to successful regional integration. In order to meet this objective, a thematic analysis is used to engage in a general exploration of the theoretical foundations and empirical settings of regional economic communities in Africa. The chapter relies on secondary data to identify themes, compare and contrast the themes, and then build theoretical conclusions from the themes explored and examined in order to address the question raised about war and conflict and the way that they relate to regional integration and its failure or success.

Regional integration–conflict/war–peace nexus

Upon the setting up of Agenda 2063, there was a debate on whether war and conflict hinder integration in Africa. This debate has been ongoing for a long time, with no consensus. Bah (2013) studied the link between civil conflict and regional integration in Africa and found that conflict prevention and resolution is a pre-requisite for regional integration. And, furthermore, suggested that peace and security top the political agenda of African regional economic communities (RECs). Ikubaje and Matlosa (2016) studied the relationship between conflict and regional integration in Africa and found that political stability is a precondition for regional cooperation. They concluded that formal processes to advance integration are usually against the challenges of war and distrust. This view is also supported by the study by Bah (2013) and the assumption of Agenda 2063.

Bah and Tapsoba (2010) examined the consequences of civil war on regional integration in Africa and concluded that intrastate conflict is a major hindrance to integration, and that the prevention and resolution of civil conflict should be given policy priority in African RECs. As observed by Healy (2011), Bah (2013), Ikubaje and Matlosa (2016), and Bah and Tapsoba (2010), the long history of conflict is a huge obstacle to the achievement of regional integration. However, Ebaye's (2010) examination of the regional integration–conflict nexus in Africa found that civil conflict had little or no relation to the success or failure of integration. The study, though, recommended that African countries should strengthen national commitment to achieving regional objectives and enhance sound civil–military relations among them, as well as commitment to democracy and human rights; rather than concentrating just on the military components of security cooperation. Voronkov (1999) too, examined the regional integration–conflict nexus and found no connection. Francis (2006) critically examined the notion that under situations of insecurity regionalism is impossible. On this point he agrees with the studies done by Voronkov (1999), Ebaye (2010) and IDRC (2009). Some scholars disagree with REI implementation approaches. For example, the Economic Commission for Africa (ECA, 2004), observed that a regional economic integration process that generates uneven and asymmetrical benefits for its member-states without adequate remedies may likely generate conflicts. This observation is supported by Chirisa et al. (2014), who established that the risk of conflict increases with progression of integration, if not handled carefully.

In general, the progress of regional integration in Africa raises many challenges and doubts on its expected success that require emergency interventions. Many scholars attempt to address some potential interventions, but there is no consensus. Their potential interventions contradict Bah (2013), Bah and Tapsoba (2010), Ikubaje and Matlosa (2016), and Healy (2011) who contend that war and conflict prevention and resolution are prerequisites for regional integration. On the other side, Ebaye (2010), Voronkov (1999), IDRC (2009) and others assume a lesser effect of war and conflict on the success of regionalism in Africa. These findings fuel active debate on the effect of war and conflict on integration; with the result that some scholars have lost their hopes on the dream of REI in Africa. A case in point is the argument by Hartzenberg (2011) that REI in Africa is in doubt, because integration efforts in African countries have not been effective in what they were established to do. While they are characterised by ambitious targets, they have a dismally poor implementation record (Hartzenberg, 2011). African regional integration arrangements are generally ambitious schemes with unrealistic time frames towards deeper integration and in some cases even political union (Hartzenberg, 2011; Chirisa et al., 2014; Nathan, 2004). Furthermore, NEPAD (2016) confirms that neither the 1980 Lagos Plan of Action (LPA)– which imagined the division of Africa into five regional economies – nor the Abuja Treaty, which replaced it in 1991, can be said to qualify as a success. Indeed, the Abuja Treaty, which proposes to implement a single currency across Africa by 2028, is quite unlikely to happen (NEPAD, 2016; Umar, 2014). The debate is ongoing and needs active intervention on whether war and conflict in Africa is a major barrier to integration or not. But currently, there is no comprehensive study done on examining this claim. This chapter attempts to fill that gap.

A historical overview of the development of African integration plans/ institutions

To understand the actual or potential causes of the success or failure of regionalism in Africa, we need to trace its history. The foundation of regional integration in Africa was laid down on 25 May 1963 in Addis Ababa, with 32 signatory governments for the establishment of a main framework of integration, the Organisation of African Unity (OAU) (UNECA, 1984). The primary aims of the OAU were to coordinate and intensify cooperation of African states in order to achieve a better life for Africans and to defend the sovereignty, territorial integrity and independence of African states (UNECA, 1984). Due to the economic pressure the continent faced in 1970s, African states adopted the OAU at its 16th Ordinary Session, held in Monrovia, Liberia, in July, 1979 (OAU, 1980). In this session the Monrovia Declaration of Commitment of Heads of State and Government of the OAU was born. The declaration laid down guidelines and measures for national and collective self-reliance in economic and social development for the establishment of a new international economic order (OAU, 1980). In adopting the declaration, urgent action was taken to establish the LPA and Final Act of Lagos (FAL) to provide political support necessary for the success of measures to achieve the goals of rapid self-reliance and self-sustaining development and economic growth (OAU, 1980).

The other parallel objectives of the LPA and FAL were the establishment of an African Economic Community (AEC) by the beginning of the twenty-first century (UNECA, 1990). In adopting the LPA and FAL it was recognised that national governments were to be responsible for implementing these strategies in their respective development plans (UNECA, 1990). However, regional bodies like the Economic Commission for Africa (ECA) would provide technical assistance to member states (UNECA, 1990). Unfortunately, very few governments made efforts to incorporate the objectives enunciated in the LPA into their development plans (UNECA, 1990). Perhaps it is worth recalling that the LPA had failed to provide for an effective monitoring and follow-up mechanism for its implementation and this may explain why governments did not feel obliged to do so.

However, the LPA and FAL went on to reaffirm the commitment to establish, by the year 2000, an African Economic Community in order to foster economic, social and cultural integration of the Africa continent (Abuja Treaty, 1991). African leaders across the continent established regional economic communities (RECs).[1] These RECs, eight in number, constitute the building blocks of the African Economic Community, with a view to unite economies of the continent as an economic community or union, to address challenges of poverty among its people, in order to remove obstacles to development in cooperative agreements in different spheres at national, sub-regional, regional and international levels (Kimunguyi, 2015; Umar, 2014; Mwithiga, 2015; Hartzenberg, 2011). The OAU transformed into the African Union (AU), which has laid out six stages of integration to be achieved over a period of 34 years and bring about the African Economic Community. These stages are laid out in Article 6 of the Abuja Treaty of 1991 (Abuja Treaty, 1991):

i.   Strengthening RECs in five years (1994–1999)
ii. Stabilisation of tariffs and other barriers to regional trade, strengthening of regional integration and infrastructure, coordination and harmonisation of RECs in eight years (2000–2007)
iii. Establishment of a free trade area and customs union at the REC level in ten years (2008–2017)
iv. Coordination and harmonisation of tariff and non-tariff between RECs in two years (2018–2019)
v.   Establishing of an African Common Market and common policies in four years (2020–2023)
vi. Integration of all sectors, establishment of an African Central Bank, African single currency – an African Economic & Monetary Union and electing the first Pan African Parliament in five Years (2024–2028).

African RECs are building blocks, whose merger will lead to the success of African integration. Based on AEC's programme, by 2011 the process of establishing a free trade area (FTA) on the continent ought to have been at an advanced stage. However, this has been delayed. Building on a policy process initiated with the LPA (1980), the adoption of the NEPAD in 2001 and the transformation of the OAU into the AU in 2002 accelerated and more strongly embedded regional integration as the transformation strategy in the continent (NEPAD, 2014). The establishment of NEPAD as Africa's blue-print for socioeconomic transformation was Africa's own resolve to put an end to the era of structural adjustment (NEPAD, 2014). In 2010, the NEPAD Secretariat was strengthened through an institutional transition to a more focused implementation agency, the NEPAD Planning and Coordinating Agency (NPCA). In view of integration into the AU structures, the NPCA strategic direction 2010–2013 was carried out, marked by the 14th AU Assembly decision of February 2010 establishing the NPCA and the Mem-orandum of July 2012 at the 19th AU Assembly approving a new organisa-tional structure for the agency (NEPAD, 2014). Africa's Agenda 2063 also builds on previous initiatives by generations of African leaders and their peoples, from the formation of the OAU, through the Monrovia Declaration in 1979, the LPA in 1980 and the Abuja Treaty in 1991, to the transformation of the OAU into the AU and establishment of NEPAD in 2001.

A brief historical overview of the institutional evolution or the development of regionalism plans in Africa like the one above helps to explain the reasons behind its failure. Despite the multiplicity of groupings, integration in Africa has not been very effective (Mwasha, 2016; Umar, 2014). The main debates among scholars revolve around why regional integration progresses very slowly, is unsatisfactory and unsuccessful. Within these debates, de Melo and Tsikata (2014), for example, state that lack of complementarities among inte-grating partners is the main cause of the failure. Fulgence (2015) and Olu-Adeyemi and Ayodele (2007) advance that lack of commitment presents the biggest challenge for successful integration. Multiple and overlapping mem-bership in RECs, poor integration models, and poor design and sequencing of integration arrangements are some of the other factors said to hinder the success of integration in Africa (Sako, 2006; Fulgence, 2015; de Melo and Tsikata, 2014; Duthie, 2011).

The ECA (2004) states that security issues in Africa hinder the success of RECs. Intra-state conflicts in Africa are said to cause political instability that reduces continental commitment for countries in conflict (Maruping, 2015). According to Jailson and Gomes (2014) and Söderbaum (1996), conflict pre-sents barriers to integration due to countries' lack of economic capacity and political will to integrate. The conflict between national and regional interests, multiple regional integration schemes, lack of clear mandate of regional schemes, and protectionist tendencies among respective African countries are generally presented as reasons for the failure of African RECs (Policy Brief, 2016; Chingono and Nakama, 2009; Hartzenberg, 2011; Söderbaum, 1996).

Qobo (2007) emphasises poor regional coordination and wrong priorities as the main reasons that harm RECs in Africa.

Mwasha (2016) observes that the individual economic problems of African countries are some of the factors which explain the failure of plans for regional integration. Most African states have suffered from harsh macroeconomic disequilibria, foreign debt service burdens, highly valued currencies, lack of trade finance and a narrow tax base, with customs duties a substantial source of revenue (Matthews, 2003). Limitation of national and regional capacities, and specifically, a lack of mechanisms and resources for an effective planning, coordination, implementation, monitoring and pragmatic adjustment of programmes on the ground have been another constraint to regional integration in Africa (Maruping, 2015). Matthews' (2003) examination of the failure of regionalism in Africa concludes that the supremacy of a few countries and a huge difference in size among members raises a question about the distribution of benefits and most regions have found it difficult to address equitable distribution of gains and losses from integration. According to Hoekman et al. (2002), the East Africa Community (EAC) collapsed in 1977 due to this problem, because Kenya had a more developed manufacturing sector than Tanzania and Uganda, resulting in large income transfers from the other two members to Kenya. Mechanisms to provide compensation to the less developed members of groupings have been either absent or ineffective in most regional economic communities in Africa. But this depends on how an economic integration scheme was designed from the very beginning (Hoekman et al., 2002). Furthermore, institutional weaknesses, including the existence of too many regional communities, a tendency towards top-heavy structures with too many political appointments, failures by governments to meet their financial obligations to regional organisations, poor preparation before meetings, and lack of follow up by sectoral ministries on decisions taken at regional meetings by Heads of States have contributed to the past failure of economic integration in Africa (Matthews, 2003). The overlapping memberships in regional communities can cause complications and inconsistencies due to conflicting obligations and divided loyalties (Okafor and Aniche, 2017; Okeke and Aniche, 2012; Aniche, 2015; Iyoha, 2005).

Africa is an economically fragmented continent and, therefore, integration should help to bring its developing countries together for mutual economic, political, cultural and social benefits (Mwasha, 2016). But in reality, the need for economic integration is usually perceived to be a result of the nature of the problems that individual African countries are confronted with in the attempts to industrialise and modernise their economies, while achieving self-sufficiency (Mwasha, 2016). These problems include "difficulties in gaining access to all required materials, following the uneven spread of natural resources and lack of funds; difficulty in finding efficient and affordable technologies to suit domestic market conditions; difficulty in securing domestic and external markets for manufactured goods" (FONDAD, 1996; Mwasha, 2016). Also, individual countries in Africa are too small to provide meaningful domestic markets for both heavy and light industrial goods produced with equipment designed for larger scales of production, thereby forcing the acceptance of ineffective production techniques (Mwasha, 2016). A large number of small countries compete with one another on international markets to sell the same agricultural products, which often reduces the strength of their bargaining on such markets, hence the need for regional arrangements to increase negotiation power (FONDAD, 1996; Mwasha, 2016).

In addition, many African economies are dependent on a narrow set of similar primary products, which generally affects their participation in world trade. Africa's participation in world trade, which has never been significant, has fallen in the last decade and intra-regional trade is itself very low (FONDAD, 1996). To offset the unfavourable trends of external markets, it is often suggested that increased trade among African nations could bring greater advantages to the nations involved and help them to mobilise their resources by finding markets for their goods. This would be especially so if it involved some regional groupings. McCarthy (1996) has observed that the small size of most of these developing economies in Africa restricts their ability to benefit from lower unit costs (derived from economies of scale) and viable import substituting opportunities, hence the argument that African countries should attempt to create an effective economic integration.

Against this background, four main objectives of the Abuja Treaty around the establishment of an economic community in Africa are:

(a) To promote economic, social and cultural development and an integra-tion of Africaneconomies in order to increase economic self-reliance and promote an endogenous and self-sustained development.

(b) To establish, on a continental scale, a framework for development, mobilisation and utilisation of human and material resources of Africa in order to achieve a self-reliant development;

(c) To promote a co-operation in all fields of human endeavour in order to raise a standard of living of African peoples, and maintain and enhance economic stability, foster close and peaceful relations among Member States and contribute to the progress, development and an economic integration of the continent; and

(d) To coordinate and harmonise policies among existing and future economic communities in order to foster a gradual establishment of the community.

The realities of regional integration in Africa: a brief review

This chapter suggests that a problem exists in the refereeing of the goals of the OAU/AU and the 1991 Abuja Treaty. This can be seen from the level of effectiveness of continental plans/institutions, starting from the LPA to FAL (1980–2000) and the Abuja Treaty. Weakness of supervision and monitoring of the goals of regional economic integration in these institutions is evident.

This observation confirms other studies conducted on related issues (see, e.g., Umar, 2014; UNECA, 1990; NEPAD, 2016). The chapter also shows that African countries place high priority on internal matters and this hinders the progress of integration. Most African countries set their policies and strategies prior to their international commitments and solution of international problems. This finding is supported by Qobo (2007) and Mwasha (2016) who observe that national policies are major obstacles for regional integration. This problem harmed the LPA and FAL, as most of the countries focused their commitments on the national level, and less or even not at all to an intra-regional and international level (UNECA, 1990).

A mismatch between national interests and priorities, and the regional interests and priorities is also observed. Kifle (1998) argues that member countries of regional groupings in Africa have individually developed their own strategies, plans and priorities while declaring their commitments to regional economic integration. This has led to a mismatch between national and regional economic policies (Kifle, 1998). In particular, Kifle (1998) contends that since independence, African countries have proposed economic integration as a desirable objective in itself as well as an indispensable means for ensuring their national economic development. However, despite several efforts, progress to that end has remained limited. A principal reason for this poor performance is that African governments have mostly followed national policies (Ricks, 2017) over and above regional ones. The desire of these countries is to achieve national objectives while they endeavour to pursue common policies towards common objectives at the same time (Kifle, 1998). This has led to the lack of harmonisation and coordination of national policies at the regional level. Even though African countries continue to speak of collective action for economic integration, no single country has as yet designed its national plans to be consistent with promoting an effective integration (Ricks, 2017). This argument seems valid also in member countries of the Inter-Governmental Authority on Development (IGAD) region, for example. Consequently, the lack of commitment to harmonise national and regional policies remains a challenge to effective integration.

This chapter also observes weak continental commitment. It is observed that a high level of international priority among the African countries distorts the strength of the continental commitment. A collective voice or force of the African continent is seen to be weak, as each country strives to achieve its own goals. This confirms other studies (de Melo and Tsikata, 2014; Olu-Adeyemi and Ayodele, 2007) that found that weak continental commitment is likely to harm the success of regionalism.

The concept of the paradox effect of war and conflict on the success of integration is also evident. This means that in relation to the success of integration, war and conflict have paradoxical effects. They do not always hinder, but sometimes support the progress of integration. Of course, conflict hinders economic growth in Africa (ECA, 2004; Söderbaum, 1996; Jailson and Gomes, 2014; Maruping, 2015). But, although war and conflict are debated by many scholars, this chapter found these factors to have both positive and negative impacts on integration progress. That is, the *paradox effect* increases or reduces the motive of a country to enter into RECs agreements. In this sense, political stability of a country is perceived to reduce the motive of a country to engage in an integration (Schiff and Winters, 2003). In other words, political instability increases the need and perceived advantages to a particular country of entering into international agreements (Schiff and Winters, 2003). This is a positive role of war or conflict on a regional economic integration process. From this theoretical standing, war and conflict have both positive and negative impacts on progress towards regional economic integration in Africa. An example of a country that demonstrates this positive impact of war and conflict on regional integration agreements is South Sudan, which joined the EAC in 2016 and the IGAD in 2011, as well as Rwanda, Burundi and Eritrea, which are all motivated to join RECs for purposes of improving their political stability.

Therefore, peace and freedom are vital factors to be maintained in a country because it creates the best avenue for establishment of RECs. But they are not guarantees of integration success. Sako (2006) emphasises that regional integration arrangements failed due to, among other things, inadequate political will and commitment to the process, high incidence of conflicts and political instability, poor design and sequencing of regional integration arrangements, multiplicity of schemes, inadequacy of funding, and exclusion of key stakeholders from regional integration processes

Regional integration in Africa: a critical theoretical analysis

Theoretical foundation of regional integration in Africa is drawn from a reference of European integration theories. Some scholars disagree on the reflection of these theories of integration in the context of Africa, on the grounds that they are drawn from or reflect European settings (Okafor and Aniche, 2017). Mitrany (1943) set forth a theory of functionalism based on certain assumptions about the causes of war and peace which stated that social and economic maladjustments are basic causes of war. He emphasised that social and economic welfare are preconditions for peace. The theory implies that a nation-state system cannot deal with basic social and economic problems because global society is arbitrarily divided into units based on territory rather than units based on problems to be solved (Mitrany, 1943). This means institutions based on function, not territory, would be appropriate for solving basic social and economic problems (Mitrany, 1943). This theory was based on a global society, and integration is driven by commonness of interests and priorities. This theory is irrelevant and invalid in African settings due to complex diversities and different development statuses.

Haas (1958) established a neo-functionalism theory that directly explained a European community. This theory of regional integration assumes that actor behaviour in a regional setting is analogous to that of a modern pluralist nation-state and takes for granted that these actors are motivated by self-interest. Haas (1958) identified political community as the terminal condition, and defined integration as a process whereby political actors in different countries are gradually persuaded to shift their loyalties, expectations and political activities towards a new larger centre whose institutions possess or demand jurisdiction over the nation-states. The theory lacks relevance to regionalism in Africa as it wholly reflects the European community that highly differs in social, economic and political settings.

Intergovernmentalism is a theory that emphasises the role of a nation-state in integration, and argues that the nation-state is not becoming obsolete due to regional integration. Milward (1992) argued that national governments of member states were primary actors in a process of regional integration, and rather than being weakened by it as some of their sovereignty was delegated to the European Union (EU), they became strengthened by the process. This is because in some policy areas, it is in member states' interest to pool sovereignty. Intergovernmentalists argue that they are able to explain periods of radical change in the EU as when interests of member state governments converge, and they have shared goals, and periods of slower integration as when the governments' preferences diverge and they cannot agree (Hoerber, 2015). They continually emphasise the role of national governments and bargaining between them in the integration process (Hoerber, 2015). This theory also reflects European settings. Key primary actors of regional integration in Africa are national governments and RECs.

These foregoing theories are not relevant to integration in Africa because they reflect the integration of highly different societies– European societies. There is a huge difference in economic status and political settings. From these theoretical misfits, this chapter supports Okafor and Aniche (2017). That is, a solution should not be sought in existing Euro-centric strategies or theories, but in a new theory; post-neo-functionalism, as suitable for the African situation and capable of solving its problems. Post-neo-functionalism advocates people-centred or human-centric or bottom-up integration rather than a top-down approach to integration, or the state-centric or inter-governmental integration of neo-functionalism. Therefore, integration should be people-driven or private-sector-led rather than state-driven or public-sector-led. The role of a state should be regulatory or facilitative (Aniche, 2014).

Regional economic integration in Africa has historical challenges to its progress that many researchers have been motivated to address. Most recent studies done on the progress of regionalism in Africa have not conclusively addressed these challenges. Okafor and Aniche (2017), Okeke and Aniche (2012), Aniche (2015) and others have examined the nature of regionalism's failure in Africa. Basically, they found that regionalism in Africa has been seriously undermined by overlapping memberships, numerous subgroups and proliferation of regional economic blocs. Their study was motivated by Ezeanyika (2006) who suggested that African countries that share similar or overlapping objectives and functions duplicate their activities when it comes to advancing regionalism, which leads to an uneconomical use of available scarce financial and human capital.

A critique of the studies by Ezeanyika (2006), Okafor and Aniche (2017), Okeke and Aniche (2012), Aniche (2015), Mwasha (2016) is that they examine intra-trade profiles by taking account of export destinations and import origin and, therefore, find less trade relationship among regional economic groups. They find that there is no strong trade relation among the member countries, and hence conclude that there is no commitment due to overlapping memberships, numerous subgroups and advancement of regional economic blocs. Their conclusion is drawn from weak methodology. Even at the highest level in Africa – that is, the Continental Free Trade Area (CFTA) level – African countries will trade more with countries outside the continent than among themselves. Low trading in intra-trade/RECs is not to be judged as ineffective, but the national trade-motive is built upon the comparative advantage theory of international trade. On the other hand, the contention that numerous subgroups and advanced regional economic blocks greatly harm progress towards regionalism is less supported and not methodologically justified fully by Okafor and Aniche (2017), Okeke and Aniche (2012), Aniche (2015).

On other considerations, overlapping of memberships and proliferation of regional economic blocs in Africa increases inter-regional trade opportunities, and enhances intra-RECs and inter-RECs integration which are important stages in regionalism progress. Overlapping memberships, numerous subgroups and proliferation of regional economic blocs increase a social and economic unity in the continent and political trust, results for minimal inter-conflict in Africa, as a relevant requirement for integration success in Africa. They break down the culturally inherited and block rigidity of RECs, as evidenced by the fact that most of them are motivated on cultural similarities. Furthermore, numerous subgroups and proliferation of regional economic blocs pave the best avenue for regionalism as it integrates or unifies African peoples.

Mwasha (2016) investigated the benefits of integration in Africa, with a focus on the EAC, and agrees that overlapping memberships due to multiplicity of economic communities hinders the integration progress. The study further shows that geographical proximity, cultural, historical and ideological similarities, competitive or complementary economic linkages, and common language among the members are desirable conditions for an effective economic integration. This means that effective integration in Africa will be achieved if Africa becomes a *monocultural continent*, which is breaking all social and cultural differences reinforced by RECs and other regional economic groups. The unity of African people is a prerequisite for successful integration in Africa. Overlapping membership positively reduces cultural and social-economic difference among the RECs, thus fostering the integration progress.

Hoekman et al. (2002, as cited by Mwasha, 2016) agrees that regional integration enhances security, but it can also worsen security in the case that distribution of transfers is asymmetrical between the member states. Mwasha

(2016) supportsHoekman et al. (2002) and contends that regional integration reduces risk of conflict in two ways. First, increasing interdependence among the members would make conflict costlier, thus, an economic integration may pave the way for political integration, substantially reducing the risk of an internal conflict. Second, regular political contract among the members can build trust and facilitate cooperation, including in conflict. A critical analysis of these findings shows that it is true that regional integration will increase both economic and political trust among the member countries, and reduces the likelihood of inter-state conflicts. In fact, regional integration can affect civil conflict, due to its limited influences of national issues of member countries. Salvatore (2004) cited by Mwasha (2016) agrees that one of the reasons for regionalism stagnation in Africa is that countries fear the loss of their sovereign powers. This means they have no will to engage in integration because they fear losing their national power. On the other hand, ignorance of private sectors in regionalism arrangements is among the factors that hinder regionalism success (Mwasha, 2016; Salvatore, 2004).

Discussion

The chapter aimed to intervene on a debate on whether or not the absence of war and conflict guarantees peace that ensures distribution, location and spatial organisation of economic activities leading to successful regional integration. Thematic facts drawn from this chapter are presented in this section. It shows that war and conflict are the *least factors* that cause the failure of African regional integration plans and have both positive and negative impacts on integration progress. Most *core factors* that hinder integration progress in Africa are outlined in the following subsections.

Problem of refereeing the goals of the Abuja Treaty

The responsibility for regional economic integration rests on RECs and individual countries (Rettig, Kamau and Muluvi, 2013). The AU Commission is charged with monitoring the continent's integration process (UNECA, 1990). This means the AU Commission is not a competent organ (referee) to preside over the goals of the Abuja Treaty and to force RECs and individual countries to enter into the regional economic integration. This problem is termed *the problem of refereeing the goals of the Abuja Treaty*. The integration process has remained slow despite numerous efforts and working committees formed by the AU to coordinate RECs, suggesting more work remains to be done. Now, many RECs have missed their target dates for implementing customs unions and common market requirements. As RECs failed to achieve integration objectives, the AU may have to play a more active role in presiding over the goals of the Abuja Treaty. The AU Commission has to extend its mandate from monitoring the regionalism implementation to a supervisory role – a competent referee that will preside over the goals of the Abuja Treaty.

High internal priorities for African countries

Most African countries are poor (World Bank, 2015). This chapter found that African countries are setting their priorities for improving social wellbeing. Priorities set involve such things as eradication of poverty by increasing school enrolments, improving healthcare, water and sanitation, infrastructure, and agricultural initiatives as the basis of family income, particularly in rural areas. Regional integration is perceived as a remote and perhaps irrelevant priority for a national interest at this stage (UNECA, 1990). It is merely assumed that regional integration has no direct influence on poverty reduction and economic growth in member states (Bundala, 2012). Apparently, there is no connection between economic growth and human development (Bundala, 2012). Taking the examples of the LPA and FAL, most countries failed to incorporate the priorities of the LPA and FAL to their national plans, because they had to prioritise high internal priorities (UNECA, 1990).

High internal priority for African countries is good for them, but hurts a collective force – the African voice – it creates egoism and individualism. African countries cannot successfully attain full integration if they are strongly bound up with their own (national) interests and commitments. High internal priority will be reduced if poverty, economic problems and other social ills in African countries are collectively reduced. The Integrated Poverty Eradication Strategy (IPES) will be the only way of involving the Collective African Initiatives (CAIs), such as shared education programmes/projects, infrastructure and health care programmes/projects, that are directly improving wellbeing of African societies. Regional economic gearing institutions such as an African Development Bank (AfBD) should play a role in financing and technical advice provision. Planning and implementation of IPES in Africa will be of multiple advantage as it is expected to increase collective voice/force, unity and reduce a mismatch between national interests and priorities, and the continental goals, and also, reduce the high internal priority of African countries.

Mismatch of national interests and priorities

Commitment by governments to incorporate regionally adopted policies into national programmes is crucial for an effective regional economic integration. However, this is not the case in Africa in general. This problem is due to the diversity of the national needs, wants and priorities of African countries. Commonness of interests, even cultures, is mostly activated by the integration progress in Africa. Taking an example of a community of the Arab Maghreb Union (AMU), it is highly motivated by culture and beliefs. From this finding, cultural differences play a role in hindering the integration progress in Africa. It is impossible to meet the interests of each country in Africa individually, but there is a need to deal with the generic African need – poverty. Poverty eradication in Africa is the first priority. Each country in Africa is struggling to eradicate poverty by any means. Integrating the goals of regionalism to member states' national poverty eradication plans could be the best strategy to raise political wills, common interests and priorities to regionalism.

Weak continental commitments

The lack of continental commitment is likely to harm regional integration progress in Africa. It is observed that a primary objective of the OAU was to unite African countries, to have a one voice – a unity. The unity is a core and a prerequisite for collective commitments. The chapter observed that unity in Africa is still weak. *Country individualism* is observed, and is persistently increasing. The continental commitment is a maximum level of unity in the continent that is the primary aim of the OAU/AU. The achievement of this level of unity or commitment will facilitate having one voice for the continent. It is at this stage that a prosperous, integrated and united Africa will arise.

Improper design of regionalism and its implementation faults

The cumulative problems of RECs in achieving regionalism are inherited in a poor or improper design of regional integration in Africa. Poor coordination, weak political will, mismatch of national interests and priorities, and others are rooted in this improper design, which results in implementation faults – deadline overruns. The African continent is highly characterised by an extreme poverty, income and social inequality (Bundala, 2017a, 2017b, 2017c). Poverty and inequality in most African countries amounts to a mismatch between national interests and the Abuja Treaty's goals; countries put more interest and prioritisation of national development plans on poverty and inequality (Bundala, 2017a). In essence, we learn that the current model of regional integration in Africa is primarily focused on economic growth rather than on social development. This became a technical design fault of regional integration plans. The model assumes economic growth, that will be accelerated by regionalism in Africa, will ensure or guarantee human development, which is likely to prove untrue. Some studies show a mismatch between economic growth and human development or poverty eradication in Africa (World Bank Group, 2015) In addressing this technical problem, the AU should seriously consider the social development of member states. That is, the regionalism will be human centred.

War and conflict in Africa

Paradoxes of the impact of war and conflict on regional integration success are shown by the fact that the presence of war and conflict in the continent may in some cases activate a motive for regional economic integration, but at the same time, in some countries war and conflict weaken or reduce the capacity of countries to access opportunities for regional economic integration. Besley and Persson (2008) established that a risk of civil conflict results in a lower investment in state fiscal capacity, notably because of a diversion of fiscal resources to increased military expenditures. However, war and conflict have bilateral effects in different situations/settings. On the other side, it increases motivation for regional integration and reduces capacity and motives of countries to engage in a regional economic integration (Tapsoba and Bah, 2010). This is the paradox of war and conflict in regard to regional economic integration.

Conclusion

The chapter explained the nature of regional economic integration failure in relation to war and conflict in Africa. The aim was to thematically investigate the interplay of regional integration success or failure, and war and conflict in Africa. The chapter lays out theoretical evidence on whether the absence of war guarantees a peace that ensures distribution, location and spatial organisation of economic activities, leading to successful or failed regional integration. The chapter established the *core factors* for the failure of REI plans in Africa. These include the problem of refereeing the goals of regional integration, high internal priority for African countries, mismatch between national interest and priority among member countries to the Abuja Treaty, weak continental commitments, high internal political stability and state freedom of the African countries. The chapter concluded that peace –the absence of war and conflict – that can be brought by the removal of firearms is important but not a guarantee for regional integration success. It can, however, create a good environment for regional integration although it may not necessarily lead to successful integration. Although peace does not guarantee successful integration, it has a positive value to it, and should be maintained. Furthermore, RECs will be used to promote peace in Africa as the chapter found RECs increase economic and political trust among African countries.

Two major policy implications arise from this. First, relying on REI and peace – the absence of war and conflict. The chapter found that regional integration promotes peace in Africa as we see some countries enter into REC arrangements for seeking peace in their countries. Moreover, integration reduces the risk of inter-state conflicts and increases both economic and political trust. Therefore, it is recommended that the AU should balance its successes in minimising African conflicts with the importance of doing more to promote economic integration. While addressing flashpoints of war and conflict is an important short-term necessity, increasing intra-African trade, building an African unity and continental commitment on integration, and networking African interdependence may offer great long-term promise (Rettig, Kamau and Muluvi, 2013). These are all steps toward the same goal of a prosperous and peaceful Africa.

The second implication is on handling the core factors of the failures of integration. While the AU does not have the authority to overcome the high internal priorities of African countries, a lack of political and continental commitments, mismatch between national interests and priority to the Abuja Treaty, or other challenges that African countries and RECs may face or bring to the table, it can and better follow its mission in encouraging integration. If the AU hopes to realise its goal of a United Africa by 2028, it must engage the continent's RECs and assist in resolving the numerous obstacles they face. It should consider expanding its efforts to coordinate regional initiatives within low-capacity countries and work to ensure that future programmes are better targeted and more visible. Further, the AU should exercise leadership to reinforce the refereeing of the goals of the Abuja Treaty in countries that do not seem to have the political will to move towards integration. It could also move from biannual meetings to more common ones and more vigorously assist in mobilising resources and coordinating their application toward regional infrastructure projects to boost trade. The AU could even consider launching voluntary international governance initiatives, such as a two-term limit for political leaders or the teaching of a common language (Rettig, Kamau and Muluvi, 2013). It can also more closely oversee and facilitate the long and difficult negotiations of protocols, and may use scorecards and penalties while monitoring their implementation to ensure that states are motivated to meet their benchmarks.

Note

1 These are: the Common Market for Eastern and Southern Africa (COMESA) formed 1993; Economic Community of West African States (ECOWAS) formed 1975; Southern African Development Community (SADC) formed 1992; East African Community (EAC) formed in 1999; Economic Community of Central African States (ECCAS) formed 1983; Intergovernmental Authority on Development (IGAD) formed in 1986; Community of Sahel-Saharan States (CEN-SAD) formed in 1998; and Arab Maghreb Union (AMU) formed in 1989.

References

Abuja Treaty. 1991. *Treaty Establishing the African Economic Community.* Abuja, Nigeria: OAU.African Union. 2005. Decisions, Declarations and Resolution. Decisions of the Assembly of Sirte Libya Assembly of the African Union Fifth Ordinary Session.

African Union Commission (AUC). 2015. Agenda 2063. The Africa We Want. A Shared Strategic Framework for Inclusive Growth and Sustainable Development. First Ten-Year Implementation Plan 2014–2023. www.un.org/en/africa/osaa/pdf/a u/agenda2063-first10yearimplementation.pdf

Aniche, E. T. 2014. Problematizing Neo-functionalism in the Search for a New Theory of African Integration: The Case of the Proposed Tripartite Free Trade Area (T-FTA) in Africa. *Developing Country Studies* 4(20): 128–142.

Aniche, E. T. 2015. The "Calculus" of Integration or Differentiation in Africa: Post-neo-functionalism and the Future of African Regional Economic Communities (RECs). *International Affairs and Global Strategy* 36: 41–52.

Aniche, E. T. 2018. Post-neo-functionalism, Pan-Africanism and Regional Integration in Africa: Prospects and Challenges of the Proposed Tripartite Free Trade Area (T-FTA). In S. O. Oloruntoba and V. Gumede (Eds), *State and development in post-Independent Africa*. Austin, TX: Pan-African University Press, pp. 155–174.

Aniche, E. T. and Ukaegbu, V. E. 2016. Structural Dependence, Vertical Integration and Regional Economic Cooperation in Africa: A Study of Southern African Development Community. *Africa Review 8*(2): 108–119.

Bah, A. 2013. Civil Conflicts as a Constraint to Regional Economic Integration in Africa. *Defence and Peace Economics* 24(6): 69–92.

Bah, A. and Tapsoba, S. J. A. 2010. *Civil Conflicts and Regional Economic Integration Outcomes in Africa*. CERDI, Etudes et Documents, E 2010.09.

Besley, T. and Persson, T. 2008. Wars and State Capacity. *Journal of the European Economic Association* 6: 522–530.

Bundala, N. N. 2012. Economic Growth and Human Development; a Link Mechanism: An Empirical Approach (MPRA Paper47648). University Library of Munich. http://mpra.ub.unimuenchen.de/47648/1/MPRA_paper_47648.pdf

Bundala, N. N. 2017a.Measurement of Social Inequalities in Developing Countries: Lessons from Africa. *The 3rd Tanzanian Association of Sociologists National Conference Proceedings* 1(1): 87–110.

Bundala, N. N. 2017a. The Poverty and Inequality Features and Determinants: A Macro Level Analysis in African Countries. *Innovative Journal of Business, Management and Economics* 1(1): 1–19.

Bundala, N. N. 2017c. Reflections on Macro-factors Associated with the Description of Poverty in the Society: A Macro-analysis of the African Context. *The 3rd Tanzanian Association of Sociologists National Conference Proceedings* 1(1): 17–38.

Chingono, M., and Nakama, S. 2009. The Challenges of Regional Integration in Southern Africa. *African Journal of Political Science and International Relations* 3(10): 396–408.

Chirisa, I. E. W., Mumba, A. and Dirwai, S. O. 2014. A Review of the Evolution and Trajectory of the African Union as an Instrument of Regional Integration. *SpringerPlus* 3: 101–121.

de Melo, J. and Tsikata, Y. 2014. *Regional Integration in Africa, Challenges and Prospects*. WIDER Working Paper 2014/037.

Duthie, S. R. 2011. *African Integration: Many Challenges, Few Solutions*. Cape Town: University of Cape Town.

Ebaye, S. E. N. 2010. Regional Integration and Conflict Management in Africa. *An International Multi-Disciplinary Journal* 4(2): 276–293.

Economic Commission for Africa (ECA). 2004. Assessing Regional Integration in Africa. ECA Policy Research Report.

Ezeanyika, S. E. 2006. *The Politics of Development Economy in the South: Problems and Prospects*. Owerri, Nigeria: Development Studies Research Group (DESREG).

FONDAD. 1996. *Regionalism and the Global Economy: The Case of Africa*. The Hague: FONDAD.

Francis, D. J. 2006. Linking Peace, Security and Developmental Regionalism: Regional Economic and Security Integration in Africa. *Journal of Peace Building & Development* 2(3): 7–20.

Fulgence, N. 2015. War on Terrorism in Africa. A Challenge for Regional Integration and Cooperation Organisation in Eastern and Western Africa. *Journal of Political Science and Public Affairs* 1: 1–11.

Haas, E. B. 1958. *The Uniting of Europe: Political, Social, and Economic Forces, 1950–1957* (3rd ed.). Notre Dame, IN: University of Notre Dame Press.

Hartzenberg, T. 2011. Regional Integration in Africa. World Trade Organisation, Economic Research and Statistics Division, Staff Working Paper, ERSD-2011–2014.

Healy, S. 2011. Hostage to Conflict:Prospects for Building Regional Economic Cooperation in the Horn of Africa. A Chatham House Report, The Royal Institute of International Affairs Chatham House, 10 St James's Square, London SW1Y 4LE.

Hoekman, B., Matoo, A. and English, P. 2002. *Development, Trade and the WTO.* Washington, DC: The World Bank.

Hoerber, T. 2015. Revisiting Alan S. Milward, The European Rescue of the Nation-State. *Journal of Contemporary European Research* 11(4): 388–391.

Human Rights Watch. 2016. World Report, 2016. Events of 2015. United States of America.

Ikubaje, J. and Matlosa, K. 2016. *The Nexus between Regional Integration and Conflicts in Africa. Department of Political Affairs.* Addis Ababa: African Union Commission.

International Development Research Centre (IDRC). 2009. Taking Stock of Post Conflict Peace Building and Charting Future Directions. Paper presented on the 10th Anniversary of Agenda for Peace, International Development Research Centre, Ottawa, Canada.

Iyoha, A. M. 2005. *Enhancing Africa's Trade: From Marginalisation to an Export-Led Approach to Development.* African Development Bank, Economic Research Working Paper Series, 77.

Jailson, E. and Gomes, G. R. 2014. The Challenges of Regional Integration in Africa. *European Scientific Journal,* special edition: 429–433.

Kifle, W. 1998. Economic Integration in Africa. *Ethioscope* 3(2), 7–12. Kimunguyi, P. 2015. Regional Integration in Africa: Prospects and Challenges for the European Union. Contemporary Europe Research Centre. Refereed paper presented to the Australasian Political Studies Association Conference, University of Newcastle, Australia, 25–27 September.

Maruping, M. 2015. *Challenges for Regional Integration in Sub-Saharan Africa: Macroeconomics Convergence and Monetary Coordination.* Africa in the World Economy. The Hague: FONDAD.

Matthews, A. 2003. *Regional Integration and Food Security in Developing Countries.* Rome: FAO.

McCarthy, C. 1996. Regional Integration: Part of the Solution or Part of the Problem? In S. Ellis (Ed.), *Africa Now.* London / Portsmouth, NH: James Currey / Heinemann, pp. 21–42.

Milward, A. S. 1992. *The European Rescue of the Nation-State* (2nd ed.). London: Routledge.

Mitrany, D. 1943. A Working Peace System: An Argument for the Functional Development of International Organization. Excerpt in Mette Eilstrup-Sangiovanni (2006). *Debates on European Integration.* London: Palgrave Macmillan, pp. 37–42.

Mwasha, O. N. 2016. The Benefits of Regional Economic Integration for Developing Countries in Africa: A Case of East African Community (EAC). *Korea Review of International Studies* 1(1): 69–92. Mwithiga, P. M. 2015. The Challenges of Regional Integration in the East Africa Community. In Adam B. Elhiraika, Allan C. K.

Mukungu and Wanjiku Nyoike (Eds), *Regional Integration and Policy Challenges in Africa*. Geneva: Springer, pp. 89–108.

Nathan, L. 2004. Mediation and the African Union's Panel of the Wise. In F. Shannon (Ed.), *Peace in Africa: Towards a Collaborative Security Regime*. Johannesburg: Institute for Global Dialogue, pp. 63–80.

NEPAD. 2014. *Strategic Plan 2014–2017*. NEPAD Planning and Coordinating Agency. https://dev.au.int/en/auc/strategic-plan-2014-2017.

NEPAD. 2016. *Regional Integration: Make Africa Win*. www.nepad.org/content/regional-integration-make-africa-win.

Okafor, J. C. and Aniche, E. T. 2017. Deconstructing Neo-functionalism in the Quest for a Paradigm Shift in African Integration: Post-neo-functionalism and the Prognostication of the Proposed Continental Free Trade Area in Africa. *IOSR Journal of Humanities and Social Sciences 22*(2): 60–72.

Okeke, V. O. S. and Aniche, E. T. 2012. Economic Regionalism and Dependency in Africa: A Study of African Economic Community (AEC). *Arabian Journal of Business and Management Review* 1(11): 5–23.

Olu-Adeyemi, L. and Ayodele, B. 2007. The Challenges of Regional Integration for Development in Africa: Problems and Prospects. *Journal of Social Science* 15(3): 213–218.

Organisation of African Unity (OAU). 1980. *Lagos Plan of Action for the Economic Development of Africa 1980–2000*. Addis Ababa: OAU.

Policy Brief. 2016. Meeting the Challenges of Regional Integration, Intra-African Trade and Economic Growth in Africa. Available as pdf.

Qobo, M. 2007. The Challenges of Regional Integration in Africa, in the Context of Globalisation and Prospect for a United State of Africa. ISS Paper 145.

Rettig, M., Kamau, A. W. and Muluvi, A. S. 2013. The African Union Can Do More to Support Regional Integration. Brookings and Trade and Foreign Policy Division, KIPPRA. www.brookings.edu/blog/up-front/2013/05/17/the-african-union-can-do-m ore-to-support-regional-integration/.

Ricks, T. 2017. From the Abuja Treaty to the Sustainable Development Goals. Realizing Economic Integration in Africa. HeinOnline. https://heinonline.org/HOL/La ndingPage?handle=hein.journals/ncjint42&div=10&id=&page=.

Sako, S. 2006. *Challenges Facing Africa's Regional Economic Communities in Capacity Building*. ACBF Occasional Papers, 5. Harare: ACBF.

Salvatore, D. 2004. *International Economics* (8th ed.). New York: Wiley International Edition.

Schiff, M. and Winters, L. 2003. *Regional Integration and Development*. Washington, DC: World Bank.

Söderbaum, F. 1996. *Handbook of Regional Organisation in Africa*. Uppsala: Nordiska Afrikainstitutet.

Tapsoba, S. J. and Bah, A. 2010. Civil Conflicts and Regional Economic Integration Outcomes in Africa. CERDI, Etudes et Documents, E 2010.09.Umar, D. J. 2014. An Evaluation of Implementation of the Abuja Treaty and the African Integration Process (2002–2012). Department of Political Science/International Studies, Faculty of Social Sciences Ahmadu Bello University, Zaria.

United Nations Economics Commission for Africa (UNECA). 1990. Conference of African Ministers of Trade Meeting11th session, 15–19 April, Addis Ababa, Ethiopia.

United Nations Economics Commission for Africa (UNECA). 1984. Conflict, Peace and Regional Economic Integration in Southern Africa: Bridging the Knowledge Gaps and

Addressing the Policy Challenges. Southern African Seminar Series, 7–8 October, Livingstone, Zambia.

Voronkov, L. 1999. Regional Cooperation: Conflict Prevention and Security through Interdependence. *International Journal of Peace Studies* 4(2): 1–8.

The World Bank. 2015. *Annual Report*. Washington, DC: World Bank.

World Bank Group. 2015. *Tanzania Mainland Poverty Assessment*. Washington, DC: World Bank.

11 Innovations for "silencing the guns" and redefining African borders as promoters of peace, wellbeing, regional and continental integration

Christopher Changwe Nshimbi and Inocent Moyo

On borders, conflict and the possibility of peace

We set out, in this book, to contribute to debates on the many and complex causes of conflict in Africa. Our focus was on the porous nature of nation-state borders. We considered the possibility that the borders could catalyse war and conflict. Hence, conditions in which, on one extreme, some African borders and borderlands constitute sites and zones of violent confrontation were examined in this book. Here, states and insurgents engage each other. Cases in point examined in the book include the Afar Horn (Chapter 3), Nigeria/Cameroon (Chapter 4), Burundi, the DRC, Rwanda and the Great Lakes and Mano River regions (Chapter 5, Chapter 6). The book has also examined what might pass for milder forms of state–state and diplomatic tensions and conflicts between states or their agencies and grassroots actors. These have been shown to play out on common transboundary resources such as rivers, lakes and river basins (Chapter 7). Thus, this book centred on the way in which the borders in Africa might exacerbate war and conflict and how this impacts on regional and continental integration. In addition, the notion that borders could be sources or facilitators of peace was discussed and the extent to which this could contribute to regional integration, peace, development and wellbeing. In this respect, the various chapters in the book went beyond the outlining, description and analysis of borders and the causes of conflict in Africa. They also considered initiatives and innovations that could be established or are already at work at grassroots, within the current integration debates and practices, to ultimately achieve development.

Peace and development

It is worth noting vis-à-vis this understanding that even conditions defined by the absence of extreme violence or war in these spaces do not necessarily translate into peace. The contributions in this book have clearly demonstrated how, though some countries may not be experiencing open war, their citizens suffer still. The suffering is observable in cross-border phenomena which manifest themselves through undocumented migration, human smuggling and, even, xenophobic attitudes towards migrants in host countries that are not necessarily at war (Chapter 8). In fact, the failure of African states to carefully craft regional migration regimes engenders irregular forms of cross-border migration. To this can be added the issue of the absence of robust frameworks within RECs, such as the Southern African Development Community (SADC). The SADC effectively conducts and guides the conduct of elections, with the result that disputed elections which yield a significant number of cross-border migrants, as happened in 2008 in Zimbabwe, is a worrying phenomenon. Although intended to establish and maintain peace, elections end up being manipulated to the point where they result in heightened levels of injustice (and violence) that induce migration of individuals who experience, among others things, absence of inner peace (Chapter 8, Chapter 9).

Concerning borders and grassroots/non-state actors

Caught within, in between and by either extreme or benign violence in borderlands or inside the territories of the AU's 55 member states, are ordinary people at the grassroots. The understanding exists that one or a combination of factors may be responsible for seeing them uprooting and migrating across borders in Africa. A classic example of these actors is the informal cross-border traders that are a common and widespread phenomenon on Africa's borders and borderlands. These actors play an important economic and social development function and are indeed responsible for a bottom-up process of regional integration (Nshimbi, 2017, 2019; Nshimbi and Moyo, 2017, 2018; Moyo, 2017, Moyo and Nshimbi 2017). The nature, dynamics and operation of these actors across African borders and borderlands provides a cross-border architecture that can be utilised for peace and development. This is primarily because of the strong, almost familial, sense of community between borderlands inhabitants. They share common history and ethnic background. Ironically, even in the Afar Horn region (Chapter 3), people at the grassroots are knitted together by overlapping institutions, relationships, identities and infrastructure that ensure peaceful and cooperative closeness, solidarity and fraternity. This is amidst state and non-state violence. It is thus paradoxical that borders and borderlands in such places as the Afar Horn are mostly understood to be sites and zones of violent confrontation which have also given way to the free reign of illicit networks and activities of insurgents. This understanding informs the general picture painted of African borders as unmanageably anarchic, insecure and inherently violent zones of insecurity.

African borders, continental integration and holistic development

Based on these discussions and findings, what conclusions can then be made about the prospects of integrating Africa and achieving holistic development? Insofar as concerns borders, it is important to note that no one-size-fits-all solutions exist. The different contributions to this volume bear clear witness to that. Cross-border or borderlands disputes in places such as the Kariba Dam (Chapter 7) have been shown to be amicably resolved and even "overlooked" by conflicting parties at the grassroots. However, where violent insurgency and conflict really exists, nation-states are on their part equally culpable of amplifying conflict and legitimising violence on some of Africa's borders and borderlands. This is because of the security stance they adopt and concerted efforts to reinforce borders.

The efforts contradict the artificiality of and the way in which these borders represent an imposed arbitrary political construction. They make the assertion hold true that non-state actors in Africa who often suffer legitimised state violence and segregation and exclusion from national wealth, such as inhabitants of the Afar Horn (Chapter 3), are subjected to the bondage of boundaries (Ndlovu-Gatsheni and Mhlanga, 2013). Whatever the case, the cardinal point to make is that, for as long as Africa continues to be bedevilled by war and conflict within and across the borders of some of the 55 sovereign member states of the AU, the prospects of successfully integrating it and achieving holistic development and the AU's ambitious Agenda 2063 will remain elusive. The cause for peace, leading to development is, however, not completely lost. Agenda 2063 itself explicitly seeks to transform Africa into an economically thriving and peaceful continent. An aspiration of this agenda notably highlights the need to prevent conflict through dialogue and to manage and resolve existing conflicts. It aims to "silence the guns" in Africa by the Year 2020 (African Union Commission, 2015). Thus, the First Ten Year Implementation Plan of Agenda 2063 stresses that African countries should end all civil and violent conflicts, wars and gender-based violence, among others, in order to collectively "silence the guns" on the continent by 2020. Though ambitious, African security experts say that programmes under this plan – such as the roadmap for dealing with arms – will help achieve peace and security on the continent (Alusala and Paneras, 2018).

Also, and very noteworthy, the welcome initiative by the AU to bring women on board represents a previously neglected and overdue innovation to establish durable peace in Africa. The internalisation of the Women, Peace and Security (WPS) agenda within the AU should break down hegemonic masculinities (Haastrup, 2017). The innovation also increases the chances of attaining peace and security because it democratises the process. It brings the process down to the grassroots. This is the space from where women operate. The same women who suffer the most from conflict but also have greater capacity to broker and maintain peace. Being on the ground, women's networks and civil society organisations tend to be the first responders to and are deeply involved in peacebuilding and efforts to restore communities in the aftermath of war and violent conflict. There lies the importance of peace and stability, for promoting positive development trajectories for regional economic communities and Africa's general development from the bottom, in borderlands and within states, up. While it is clearly understood that peace is not the only influential factor in this respect, it is imperative that a general state of peace across Africa is attained if meaningful socioeconomic progress and political advancement are to be guaranteed.

Regional integration is also assumed to be critical to positively contributing to ending war and preventing conflict. This goal is not totally new or unachievable. African countries and others elsewhere in the world have promoted cross-border cooperation in some sectors of the economy. This has eventually fostered integration between those countries and spilled over into other areas of cooperation, furthering integration and entrenching peace. Such a model of cooperation, leading to greater integration and peace is recommended for some African countries which seemingly are unable to cooperate for mutual benefit but display a propensity to dispute over common natural resources (Chapter 7). However, this book has shown that even where ungoverned spaces between proximate states exist and provide a conducive environment for violent insurgency that contributes to state fragility (Chapter 4), rising unusual security concerns that spill over into neighbouring countries can foster cooperation between the governments of affected neighbouring countries. For instance, following a 2014 summit in which the Presidents of Benin, Cameroon, Chad, Niger and Nigeria agreed to coordinate and jointly combat insurgent Boko Haram, their countries constituted a counter insurgency coalition which engaged in a military campaign against Boko Haram across their common borders and borderlands (Faul, 2015). Whether this campaign has succeeded or failed in achieving its mandate is beyond the scope of this chapter. It is only cited to point to practical initiatives taken by states to address some of the challenges associated with Africa's porous borders. It also highlights the necessity of proactive strategies for cross-border cooperation, peace, wellbeing and successful integration, as was proposed in the cases of countries that share transboundary natural resources such as water. Such strategies should be designed in such ways that work towards the deliberate creation of peace through forms of cooperation across nation-state borders; that they create a conducive environment for peace itself, wellbeing and functioning regional and continental integration processes in Africa.

Through the AU, African states would indeed do well to reimagine institutional mechanisms at various levels down to the grassroots to enable them to adopt and take advantage of strategies and policies that take local realities and existing capabilities for achieving peace in those localities. It is high time states reoriented their understanding and approached borders and borderlands from a different angle than as sites and spaces that must be controlled and managed. Such a rethink is possible if constructions of the "bondage of boundaries" were dismantled based on the understanding that for the most part, African borderlands really constitute clusters of communities of similar people who were crossed (and not the other way around) (Lamb, 2014; Moyo, 2016) by the colonially constructed border. Of course, such an understanding is, simultaneously, not so reckless as to ignore present realities in which insurgencies, for instance, exist and their activities facilitated by porous borders. Rather, it appreciates the need to adopt borderlands' community-based approaches in which grassroots actors (such as the women cited above) are deeply involved and incorporated in processes of peace and security.

Differently stated, while cognisance is taken that colonial gerrymandering led to the erection of borders and the construction of others out of people who were and are still one people, the so called post-colonial states (if they are truly post-colonial or making attempts towards that goal) have no business mimicking the erstwhile colonial masters by criminalising cross-border movement. This is where the notion of governing cross-border movement and architecture for regional integration, consolidation of a Pan African identity as Africa moves towards an African Economic Community (AEC) in 2028 and a united Africa in 2063, comes to the fore. Hence, we argue that although borders in some cases start, exacerbate and export conflict and war, they can also be used as an architecture to create peace. For which case there is need to move beyond a pessimist and Eurocentric narrative of African countries torn by war from borders.

References

African Union Commission. 2015. Agenda 2063. The Africa We Want. A Shared Strategic Framework for Inclusive Growth and Sustainable Development. First Ten-Year Implementation Plan 2014–2023. www.un.org/en/africa/osaa/pdf/au/agenda 2063-first10yearimplementation.pdf (accessed 9 May 2018).

Alusala, N. and Paneras, R. 2018. Silencing the Guns by 2020 – Ambitious but Essential. *ISS Today*, 14 March. https://issafrica.org/iss-today/silencing-th e-guns-by-2020-ambitious-but-essential (accessed 9 May 2018).

Faul, M. 2015. Nigeria Postpones Elections, Focuses on Major Offensive Against Boko Haram. *The Christian Science Monitor.* 7 February. www.csmonitor.com/ World/Latest-News-Wires/2015/0207/Nigeria-postpones-elections-focuses-on-major-offensive-against-Boko-Haram (accessed 9 May 2018).

Haastrup, T. 2017. "Silencing the Guns" as Militarisation: A Feminist Perspective on African Security Practices. Paper presented at the Workshop on African Security, Uppsala, 22–24 November 2017.

Lamb, V. 2014. "Where is the Border?" Villagers, Environmental Consultants and the "Work" of the Thai–Burma Border. *Political Geography* 40: 1–12.

Moyo, I. 2016. The Beitbridge–Mussina Interface: Towards Flexible Citizenship, Sovereignty and Territoriality at the Border. *Journal of Borderlands Studies.* DOI doi:10.1080/08865655.2016.1188666.

Moyo, I. 2017. Zimbabwean Cross-Border Traders in Botswana and South Africa: Perspectives on SADC Regional Integration. In C. C. Nshimbi and I. Moyo (Eds), *Migration, Cross-border Trade and Development in Africa: Exploring the Role of Non-state Actors in the SADC Region.* Palgrave Studies of Sustainable Business in Africa. London: Palgrave, pp. 43–62.

Moyo, I. and Nshimbi, C.C. 2017. Cross Border Movements and Trade: A Tenacious Lasting Reality in the Southern African Development Community (SADC) Region. In C. C. Nshimbi and I. Moyo (Eds), *Migration, Cross-border Trade and Development in Africa: Exploring the Role of Non-state Actors in the SADC Region.* Palgrave Studies of Sustainable Business in Africa. London: Palgrave, pp. 191–208.

Nshimbi, C.C. 2017. The Human Side of Regions: Informal Cross-border Traders in the Zambia–Malawi–Mozambique Growth Triangle and Prospects for Integrating Southern Africa. *Journal of Borderlands Studies.* DOI   doi:10.1080/08865655.2017.1390689.

Nshimbi, C. C. 2019. Life in the Fringes: Economic and Sociocultural Practices in the Zambia–Malawi–Mozambique Borderlands in Comparative Perspective. *Journal of Borderlands Studies.* 34(1), 47-70.

Nshimbi, C. C. and Moyo, I. 2017. History, Trends and Dynamics of Cross Border Movements and Trade in the Southern African Development Community (SADC) Region. In C. C. Nshimbi and I. Moyo (Eds), *Migration, Cross-border Trade and Development in Africa: Exploring the Role of Non-state Actors in the SADC Region.* Palgrave Studies of Sustainable Business in Africa. London: Palgrave, pp. 1–14.

Nshimbi, C. C. and Moyo, I. 2018. Informal Immigrant Traders in Johannesburg: The Scorned Cornerstone in the Southern African Development Community Integration Project. InA. Adeniran and L. Ikuteyijo (Eds), *Africa Now! Emerging Issues and Alternative Perspectives.* London: Palgrave Macmillan, pp. 387–413.

Ndlovu-Gatsheni, S. J. and Mhlanga, B. 2013. Introduction: Borders, Identities, the "Northern Problem" and Ethno-Futures in Postcolonial Africa. In Sabelo J. Ndlovu-Gatsheni and Brilliant Mhlanga (Eds), *Bondage of Boundaries and Identity Politics in Postcolonial Africa: The "Northern Problem" and Ethno-Futures.* Pretoria: Africa Institute of South Africa, pp. 1–23.

Index

Printed in the United States
by Baker & Taylor Publisher Services

Printed in the United States
by Baker & Taylor Publisher Services